The Digital Cell

Cell Biology as a Data Science

The code associated with this book is available at
http://doi.org/10.5281/zenodo.2643411

The Digital Cell

Cell Biology as a Data Science

Stephen J. Royle

University of Warwick

COLD SPRING HARBOR LABORATORY PRESS

Cold Spring Harbor, New York • www.cshlpress.org

The Digital Cell
Cell Biology as a Data Science

Publisher	John R. Inglis
Acquisition Editor	Richard Sever
Director of Editorial Development	Jan Argentine
Project Manager	Inez Sialiano
Permissions Coordinator	Carol Brown
Director of Publication Services	Linda Sussman
Production Editor	Kathleen Bubbeo, Joanne McFadden
Production Manager	Denise Weiss
Cover Designer	Pete Jeffs

Library of Congress Cataloging-in-Publication Data

Names: Royle, Stephen J., author.
Title: The digital cell : cell biology as a data science / Stephen J. Royle,
   University of Warwick.
Description: Cold Spring Harbor, New York : Cold Spring Harbor Laboratory
   Press, [2019] | Includes bibliographical references and index. |
   Summary: "Cell biology is becoming an increasingly quantitative field, as technical
   advances mean researchers now routinely capture vast amounts of data. This
   handbook is an essential guide to the computational approaches, image processing
   and analysis techniques, and basic programming skills that are now part of the skill
   set of anyone working in the field" -- Provided by publisher.
Identifiers: LCCN 2019024022 (print) | LCCN 2019024023 (ebook) | ISBN 9781621822783
   (hardcover) | ISBN 9781621822806 (epub) | ISBN 9781621822813 (mobi)
Subjects: LCSH: Cytology--Data processing.
Classification: LCC QH585.5.D38 R69 2019 (print) | LCC QH585.5.D38 (ebook) |
   DDC 572--dc23
LC record available at https://lccn.loc.gov/2019024022
LC ebook record available at https://lccn.loc.gov/2019024023

10 9 8 7 6 5 4 3 2 1

For a complete catalog of all Cold Spring Harbor Laboratory Press publications, visit our website at
www.cshlpress.org.

To Jen, Fin, and Kay

Contents

Preface

If you are a cell biologist, you will have noticed the change in emphasis in our field. At one time, cell biology papers were—in the main—qualitative. Micrographs were of "representative cells," western blots were from a "typical experiment," etc. This descriptive style has given way to approaches that are more quantitative. Qualitative observations that were once taken at face value must now be measured and objectively assessed. More recently, as technology has advanced, computing power has increased, and data sets have become more complex, we have seen larger-scale analysis, modeling, and automation begin to take center stage. This is digital cell biology.

This change encompasses several approaches, which include (in no particular order):

- Statistical analysis
- Image analysis
- Coding
- Automation allowing analysis at scale
- Reproducibility
- Version control
- Data storage, archiving, and accessibility
- Electronic lab notebooks

These approaches are not new to biology. In fact, some fields have used them extensively for years. Perhaps most obviously, groups that identified themselves as "systems biologists" or "computational biologists" and those working on large-scale cell biology projects were early adopters. But these approaches have now permeated mainstream cell biology to such an extent that any research group wanting to do cell biology in the future must be well-versed in them in order to progress. The shift is changing the skill set that we look for when recruiting scientists now, and it will shape the cell biologists of the future. Fields such as biophysics and neuroscience are further through the change, whereas others have yet to begin. It is happening to all of biology, and it is an exciting time to be involved in research.

The aim of this book is to equip cell biologists for this change: to become digital cell biologists. Maybe you are a new student starting your first cell biology project.

This book is designed to help you. Perhaps you are working in cell biology already but you have not had much previous exposure to computer science, mathematics, and statistics. This book will get you started. Maybe you are a seasoned cell biologist. You read the latest papers and wonder how you could apply those quantitative approaches in your lab. You may even have digital cell biologists in your group already and want to know how they think and how you can best support them. There is something for you in these pages. Let's get digital.

Acknowledgments

M any thanks go to the folks in my research group who made suggestions for what to cover in the book and also contributed microscopy images. A number of people commented on various sections. I am very grateful to Julia Brettschneider, Gabrielle Larocque, Erick Martins Ratamero, Claire Mitchell, and Ellis Ryan for their help and input. I discussed the content of the book with many cell biology colleagues, including Julie Welburn and Andrew Peden, who gave useful suggestions on what to include. David Stephens, Nicola Stevenson, and Richard Sever provided very useful feedback on the content. A big thank you to my family for their patience and support with this project, especially Jen, who read the first complete draft.

This project started life as a feature on my data analysis website *quantixed*. Richard Sever had the insight to turn it into a book; my thanks also go to him for the idea.

CHAPTER 1

The Digital Cell Philosophy

The philosophy behind *The Digital Cell* is that *quantification* is the key to understanding cell biology. Results should be measurable, and any qualitative results should be quantified. Next, all analysis should be *automated* as far as possible. This is to remove bias with the aim of making analyses reproducible. To make all of this happen, *organization* is crucial so that links can be made between experimental design, execution, data, quantification, and results.

Experimental data are captured and stored in an organized way that allows easy access and reuse. Computer programs perform analysis of the raw data via a workflow or pipeline. The output is a readout of the analysis. This is the result and it is disposable. I will explain why, but first we need to clarify what we mean by a workflow or pipeline.

WORKFLOWS AND PIPELINES

Setting up an automated workflow or pipeline means that your analysis will be reproducible. Ideally, the number of steps involving a human is minimized because, sadly, it is we humans who introduce errors into analyses. Workflows and pipelines are also a huge time-saver. A manual workflow will take time. Imagine you get to the end and then decide a parameter needs to change; you want to measure something slightly differently or maybe you collect a new data set. This means you have to do it all again. If you have programmed a computer to do all of the intervening steps, all you have to do is click "go" once more.

Computational methods for processing data are referred to as workflows or pipelines. The two are similar in that you put raw data in at one end and get a final analysis out at the other. A pipeline is where this processing is seamless. The data are fed in and then automatically passed to other modules to achieve the final analysis, requiring no human intervention. A workflow, on the other hand, is where there are intermediate products that must be manually fed into another program or module before the final result is achieved. For example, a pipeline might take a directory of images, extract information, do some calculations, and make a plot of the data all within one software package. A workflow, on the other hand, might involve analyzing a directory of images using a script and then feeding the output of that analysis into a separate package to produce the final plot.

Two things are implicit in the design of a workflow or a pipeline. First, raw data are treated as read-only (unchangeable). Second, outputs (graphs, figures, and so on) are disposable. The data can be fed into the workflow again and again and again. And then fed into some other workflow in the future. This means that the data stay untouched, in the raw form in which they were captured. The raw data are a separate entity from the workflow, and any other data set should be able to be fed in. This also means that what comes out is not precious. The resulting plots and files should be designed to be updated, overwritten, or otherwise just deleted. Can't remember if the outputs were done with the latest version of the workflow? No problem! Just delete them and run it again.

This means that all anyone would ever need to reproduce your analysis is a data set and the workflow/pipeline. Do not forget that "anyone" includes you! Your future self is the immediate beneficiary of organization, documentation, automation, and reproducibility.

USING SPREADSHEETS FOR EXPERIMENTAL DATA

Developing an analysis workflow or pipeline means passing data from one program to another. These files are normally long lists of numbers or words—for example, the results of analyzing the intensity of fluorescence in several cells over many time points. Your first instinct might be to put these results into a spreadsheet program so that you can look at them. Passing data from one program to another is best achieved using simple files organized for computers (and not humans) to read them. Ideally, these files are text files of comma-separated values (csv) or some other simple format. Spreadsheet programs can create these simple format files, but it is best to avoid using such programs if possible.

Spreadsheet programs such as Microsoft Excel are incredibly easy-to-use and are ubiquitous, but they are problematic for use in science. Excel is designed for compiling sales figures in a business environment. It performs poorly as a scientific application for several reasons:

- It is not auditable. Errors can be easily introduced by accident, and the user would never know. It is very difficult to find mistakes because there is no history window allowing the user to see what has happened.
- It is not good for biological data. For example, protein names such as OCT4 or SEPT9 get automatically converted to dates.
- Worksheets are limited to 1,048,576 rows and 16,384 columns.

- Excel cannot make high-quality graphics for publication.
- Users tend to compile summary statistics in the same sheet as the primary data, which prevents simple output of raw data for analysis.
- Excel lends itself to presentation of data for humans to look at but not for the organization of data frames that a computer can readily understand.

Having said all this, avoiding Excel entirely is not practical and there are areas in which Excel does indeed excel:

- It is very useful for organizing data quickly.
- It is great for a quick visual scan of the entire data set and for spotting obvious errors.
- Excel is useful for quick calculations and for preparing simple charts to show informally.
- The charts are dynamic and update on-the-fly.
- Powerful functions like VLOOKUP() and Pivot Tables are easy to perform, whereas they can be cumbersome to perform in other packages.
- Exporting as csv or other formats for use in another program is simple.

You need to be aware of the limitations of Excel and always organize your data with future reuse in mind:[1]

- Turn off the automatic functions that you do not need, because they might interfere with your data.
- Enter data consistently. For example, stick to YYYY-MM-DD for dates.
- Do not leave empty cells. They could mean data not collected, a mistake in entry, or 0.
- Do not do calculations among the raw data; if you want to do this, use a separate worksheet.
- Do not use coloring of cells or other formatting tricks to encode information; the information cannot be exported easily.
- Organize the data in a rectangle starting at A1.
- Save out a copy as a csv file or other delimited text file.

Outputs from image analysis software are typically in csv format. So long as the data are well organized and you use a sensible filenaming system (see Chapter 2), these files will be all you need to analyze and present your data. With some practice you can avoid using spreadsheet programs entirely.

SOFTWARE FOR DIGITAL CELL BIOLOGY

In the book we will use two software types: ImageJ for image processing and R for number crunching. Specifically we will use Fiji and RStudio as environments for ImageJ and R, respectively. These environments are explained in detail in Chapter 3. They were selected because they are freely available, open-source, widely adopted, and likely to be around for a long time. We will also use the command line of your computer to perform powerful operations and unlock a new world of possibilities for you and your research. There are a number of scripts, macros, and code examples in the book.[2] These have been kept deliberately simple so that they can be understood. The aim is for you to build on these examples for your own work. To save rekeying the examples, they are available at https://doi.org/10.5281/zenodo.2643410.

FOCUSING ON IMAGING

Cell biology is a broad subject and encompasses many techniques—from structural biology and biochemistry to immunology and genetic analyses. It is not possible to cover in one short book all the data types and methods that you might use as a cell biologist. Instead I have concentrated mainly on imaging data from microscopy experiments. Microscopy is at the core of most cell biological studies, and I have chosen to concentrate on fluorescence microscopy rather than other types of microscopy data—brightfield micrographs, electron micrographs, atomic-force microscopy, single-molecule localization microscopy, and others. We will briefly look at the analysis of gels and blots, but other cell biology data types (flow cytometry, proteomics, gene expression analyses, etc.) do not receive further attention. Analyses involving these types of data have many things in common with the approaches described in the book (i.e., experimental design, unbiased analysis, statistics, reproducibility, and presentation).

GOLDEN RULES

Throughout the book, you will find "golden rules" to follow. Here are the golden rules to being a digital cell biologist.

◆ *Golden Rules*

- Quantification is key to understanding cell biology.
- Automate analyses wherever possible to minimize errors and bias introduced by humans.
- Aim for reproducible research to help the future you if no one else.
- Raw data are read-only.
- Outputs are disposable.

CHAPTER 2

Dealing with Data

Whatever your project is about, you will generate data—and lots of it. You need strategies to deal with this data or it will overwhelm you. You need to get organized. Perhaps your lab already has policies and practices in place. In which case, buy into them; they are there to help you. If you are starting from scratch, this section deals with all the things you must consider when setting up a data management plan for yourself or for your group.

WHY IS ORGANIZATION SO IMPORTANT?

Forget any pictures you may have seen of Einstein's messy desk; the most productive scientists nowadays are well-organized. Being organized means being able to find your data quickly, even if it is years old. We have a responsibility to do this for two reasons. First, we are required to hold our raw data for a period of time after publication (normally 10 years). A question may arise about one of our papers at any time, and we need our data to be organized to respond promptly to any queries about our work. Second, being organized is actually part of your job as a scientist. Your research and possibly your salary are likely paid by a grant from a funding body (charity or government organization). This transaction means that you are responsible for generating results. If there is not an adequate record of these results, then the funder has paid some money and got nothing in return. This probably breaches the terms of the grant. All data, even negative findings, must be documented and auditable.

Besides these very serious responsibilities, being organized helps the future *you*. It will help you work more efficiently and net more results promptly. Sometimes experimental data make no sense at the time they are collected. Later on, in light of new information, it can turn out that your old data now make sense. If this happens, it may be possible to publish your data, but this can only happen if your data can be accessed and understood.

The other scenario is that you cannot continue because of unforeseen circumstances, and someone else needs to carry on your important work. I have not found this scenario to be a good motivator to convince lab members to be well-organized!

They are mostly too traumatized imagining the unforeseen circumstances to worry about their scientific legacy. However, the work you do in the lab *is* important. If it wasn't, you wouldn't be doing it. So it must be preserved. Even without a death or accident, at some point you will leave the lab and somebody else will have to access your data to build on your work or to help prepare it for publication. Writing up work after somebody has left the lab, even if you are still in contact, can be very difficult or even impossible if the person has not been well-organized. In the worst case, the work simply has to be redone by someone else and your authorship on the paper is at risk.

Do not fall into the trap of believing that you will be able to just "remember stuff"—for example, details of certain experiments or which experiments were done and when. Our memories fade fast and can play tricks on us.[3] The only solution is to get organized and to document your work well.

GETTING ORGANIZED

Experiment-Based Organization

Base your entire data organization system around your experiments. To do this, come up with an accession numbering system to catalog your experiments and use this to link to your raw data, your analyses, your lab book, and so on. A simple way to do this is to use your initials and a three-digit number to catalog each experiment. Next, give your experiment a title. This helps to clarify what you are trying to do and what you want to find out.

Example

```
Accession Number: SJR089
Title: Test if GFP-protein-X leaves the plasma membrane upon
treatment with drug Y
```

If it feels like it is too difficult to give your experiment a concise title, this could be a sign that you are actually planning two or more different experiments. Using the example above, suppose that while treating cells with drug Y you would also like to find out if the actin cytoskeleton becomes altered. Now coming up with a concise title is difficult. This is because there are two different aims. You need to separate them and assign distinct accession numbers; of course, you can plan to run them in parallel, but they need to be organized separately. If you struggle to give your experiment a title at all, this might be because there is no clear aim to your experiment. Maybe you do not need to do it!

An organization that is centered around the experiment works well because it does not rely on dates, which always need to be cross-referenced. In cell biology, most experiments run over several days, and because of this, experiments are over-lapped so that you can get more done. What this means is that dates are not unique to an experiment, and so any organization based on dates gets murky. In addition, we often return to experimental results at a later date and do more analysis. With an organization system based on accession numbers, these later analyses can be easily grouped with the original work, whereas a date-based system would be discon-nected. I do not advocate getting rid of dates entirely. On the contrary, everything needs to be date-stamped to help you to track down the correct files. It's just that the *organization* of the system needs to be based around experimental accession numbers.

You need a place to store your raw data, and this needs to be organized properly. Ideally you have a *share* on a server that is backed up regularly. A share (or network share) is your allocation on a server that you can access over the network. Without getting too technical, you need to make sure that there is enough space on the server for what you will generate. A microscopy experiment can produce 200 MB to 10 GB of data. New technology is pushing these sizes even larger. You will do many experi-ments, perhaps over many years. Do you have enough space? A key step in your data management plan is to consider your storage requirements before you start acquiring data. Nowadays, most funders want to inspect your data management plan, and it is worth thinking about this carefully. Storage space needs to be continually monitored as your project progresses to ensure the storage is expanded as necessary. Data stor-age works best if all of your files are in one place, your share on the server, and not split across many locations and definitely not on portable hard drives. For more infor-mation see the subsection Backing Up Your Data.

Use your experimental accession numbering system to organize your files in your network share. In the example below, Data and Analysis are separate directories in your share on the server.

Example organization

```
1    your_share/data/imaging/SJR089_2018-08-23/
2    your_share/data/western_blotting/SJR089_2018-08-25/
3    your_share/analysis/SJR089/
```

In the example above, we did some live-cell imaging to visualize GFP-protein-X leaving the plasma membrane in response to the drug. In parallel, we made some lysates to look at GFP-protein-X expression and analyzed them by western blotting a couple of days later. We then analyzed the live-cell imaging data and put this in the appropriate folder in the analysis directory. All of the data can be easily located using the experimental accession number, in this case SJR089.

To help navigate your network share when you have many files and folders, you can make a dump of all filenames in it using the command line (Terminal on Mac, Command Prompt on Windows). See the subsection on Mastering the Command Line in Chapter 6 for more information.

```
1   cd your_share/
2   find -L . > ~/Desktop/all.txt
```

Or you can search for files corresponding to the experiment:

```
1   cd your_share/
2   find . -name '*SJR089*' > ~/Desktop/SJR089_files.txt
```

This second example will find files and folders containing the characters SJR089 in their name, and print them to a file on your desktop called SJR089_files.txt. Note that the name flag is case-sensitive; use iname if you would like to do a case-insensitive search.

When naming files and folders, get into the habit of using only alphanumeric characters in filenames and avoid using spaces. Spaces have a special meaning on the command line and need to be "escaped." Spaces can be avoided by using camel case (LikeThis) or mimicked with an underscore (like_this). These conventions make subsequent programming and analysis much more simple. Also, pick a format for dates and stick to it. The standard way of writing dates (ISO 8601) is YYYY-MM-DD (e.g., New Year's Eve in 2022 is 2022-12-31). Shorter variants are OK (e.g., 221231), but altering the order of years, months, and days causes confusion and errors. Consistency is key to organization. If you work with a protein that has several names, use only one. Cell lines should be referred to by their proper name only and so on. These details can make the difference between being able to find some files and not. Finally, including information about the file is best done in the metadata rather than in the filename. If the choice is capturing the information in the filename or not at all, then go ahead. Filenames can be long nowadays, but bear in mind that parsing long filenames with a computer during analysis can be tricky. More so when they have been generated by a human, because longer filenames mean a higher probability of typos.

Databases for Resources

Most of the work you do will involve many different reagents that you have generated. These might include DNA plasmids, cell lines, antibodies, siRNAs, oligonucleotides, and proteins. These reagents are lab resources. They cost time and money to make. You benefit from them and will contribute to them during your time in the lab. It is to everyone's benefit that these are organized efficiently. Ideally there should be a central database for each resource that allows you to quickly locate the correct reagent. Using DNA plasmids as an example, an accession numbering system is a good idea so that each DNA construct has a unique number that reveals in which box it is stored and at what position. This accession number can be used for storing different forms of the plasmid. For example, for each plasmid you might have three DNA tubes (for transfecting into cells, for cloning, and for emergencies) and a bacterial glycerol stock. All four can be stored according to this position information in the accession number. Plasmids only get deposited with an accession number once they have been fully verified. This numbering system is also useful when planning out experiments, because plasmid names can be very long and the accession number is easier to write down. Similar organization systems can be set up for other types of communal reagents. Besides reagents that you have generated, chemicals and drugs ordered direct from a supplier need organization so that they are stored correctly and can be located quickly.

Whatever system is in place in the lab, you should embrace it and contribute to it. If there is not a system in place, begin by starting your own. You will need it to organize your own reagents if nothing else. It is important for reproducibility to know what reagent you used for each experiment. Database software such as FileMaker Pro is ideal for creating robust databases for lab environments. Small or simple databases can even be built in a spreadsheet program like Microsoft Excel.

All research groups accumulate a treasure trove of experimental protocols that "just work." These should be collated into a lab manual that acts as a database for standard operating procedures. Each protocol can again have an accession number, or it might be sufficient to refer to it by name. Just like reagent databases, after putting in the effort to get a new method working, you should write up your protocol and add it to the manual; it is a precious new resource. This saves everyone time and effort, but, importantly, lab notebook entries can simply refer to the method in the manual by accession number and only detail any variation from the established protocol. So with accession numbering systems in place for important reagents and for your experiments and methods, you next need to accurately document everything you do in your lab notebook.

Electronic Lab Notebooks (ELNs)

The days of the paper lab book are numbered. Although paper lab notebooks are still widely used, the future is electronic. Electronic lab notebooks (ELNs) are superior to paper because they allow content to be searched, the entries are easy to read, they are easily accessible, and backup means they will never get lost or damaged. There are many ELN options available, and you need to ensure that the solution you choose is compatible with the requirements of your funder and your institution. There are some important issues to consider when selecting an ELN platform:

- Is it easy-to-use? Can pictures and notes be added easily?
- For versioning, are entries and edits time-stamped? Are they tamper-proof?
- Where are my notes stored? Who can read them? Consider backup and data security.
- Can I easily take my notes away? You may need to do this if the ELN platform company changes its policies or goes out of business.
- What are the costs associated with using the platform—for an individual and for the whole lab?
- Will the software be kept up-to-date?
- Is it compliant with GLP (good laboratory practice)? For example, does it allow access for the analyst, manager, and QA reviewer?
- Is intellectual property protected?

A simple solution that I favor is to self-host an ELN on a local network. We use a self-hosted version of WordPress, but other blogging platforms and even wikis have similar functionality and are in use in many labs and research institutes around the world. If you already have the hardware, it is free to set up and very easy to use. WordPress passes all the concerns above—for example, any edits are visible and changes can be rolled back if required. The major plus is that you host your own data, which counters concerns about IP or about a third party accessing your data. Self-hosting is also the biggest disadvantage: Backup and data security are your responsibility. However, with a little technical know-how, this is easily solved with plug-ins to do the backup and proper control of log-ins to the platform. There are plenty of guides online to walk through setting one up. A simple setup would be a network attached storage device (NAS box) running Linux. The NAS box needs a WordPress installation, phpMyAdmin setup, and a registered mySQL database. A single WordPress install can be run for one lab or can be configured as a multisite WordPress setup, which allows several labs to each have their own "site." A typical setup for one research group on a multisite setup would be for each lab member to be able to view all ELN entries for their own lab but not those of other labs. Each lab member can write and edit their own lab book entries but they cannot delete them. The PI or a designated admin has higher privileges and can control permissions, for example. Only registered users can log in and view anything, so the contents of the ELN are secure.

Figure 2.1 shows a typical ELN entry. Lab members can edit the entry in draft form before "publishing" it to the site. After publishing, everyone in the lab can read it. While in draft form, only the author and the PI can read it. How the ELN appears to the reader is controlled by "Themes," which can be customized.

There are other ELN solutions available, including (at the time of writing) RSpace, eLabFTW, Benchling, Findings, DEVONthink, LabCollector, and labfolder. A major concern with these packages is how well they will be maintained in the future. WordPress is ubiquitous, and we can be confident that it will be around for many years. More commercial options include Microsoft OneNote, Apple Notes, and Evernote. These are very powerful bits of software, used by many, but separating a lab member's work entries from their personal stuff creates a headache.

The major benefit is that it is very simple to find information in an ELN via search or through the use of categories and tags. Templates can be set up so that entries are much faster to complete and they will contain more information than a written record would do.

ELNs are well-suited to the data organization described so far. ELN entries should have the experimental accession number in the title followed by the title of the experiment itself. Any reagents used can be referred to by their accession number, and any data or subsequent analysis can be inserted or linked. This makes the ELN entry the way to decode an entire experiment from start to finish (Fig. 2.2).

Databases for Imaging Data

For imaging data, there are open database solutions to help you to organize your work. The current leading software platform is the Open Microscopy Environment (OME) server and OME Remote Objects (OMERO), which allow access to your data stored on a server in raw form. All the metadata associated with the experiment are preserved. Analysis and the ability to make figures can happen within OMERO, and plug-ins allow it to talk to image analysis software, such as ImageJ (a widely used image analysis program; see Chapter 4). Images can be organized into *projects* and *data sets*. What works well is to have few projects and many data sets. A suggested organization is to have one project for each big topic you are working on. For example, your main project might be on mitochondrial transport on microtubules. If you

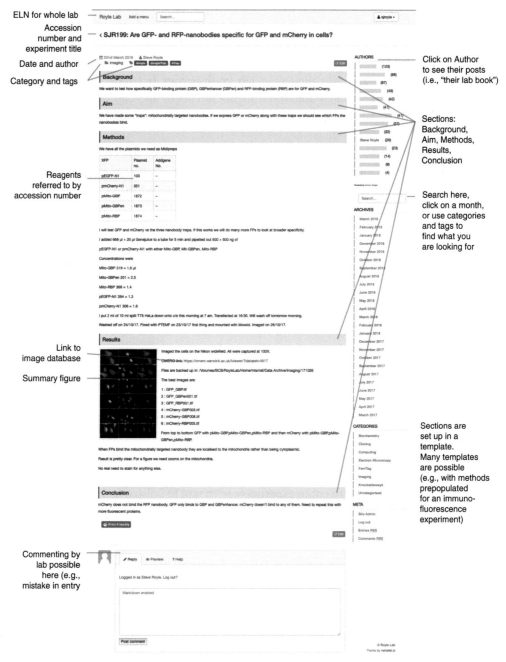

ELN for whole lab

Accession number and experiment title

Date and author

Category and tags

Click on Author to see their posts (i.e., "their lab book")

Sections: Background, Aim, Methods, Results, Conclusion

Reagents referred to by accession number

Search here, click on a month, or use categories and tags to find what you are looking for

Link to image database

Summary figure

Sections are set up in a template. Many templates are possible (e.g., with methods prepopulated for an immuno-fluorescence experiment)

Commenting by lab possible here (e.g., mistake in entry

FIGURE 2.1. An example ELN entry.
A completed record has an experimental accession number and title. Assigning a category and adding tags further aid searchability. We use a template with the following sections that can be edited. Background: What happened before? Aim: What did you want to test? Methods: What did you do? Results: What happened? Conclusion: What did you find out? What does it mean? The theme shown here is a modified form of Gitsta.

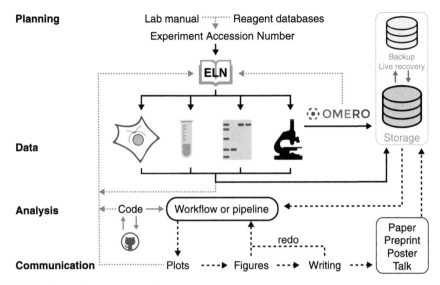

FIGURE 2.2. Getting organized.
Planning: An experiment is planned using information in the lab manual and lab reagent databases. It is given an accession number, and this is used to locate everything related to the experiment. The electronic lab notebook (ELN) is at the center of this organization.
Data: Experimental results are acquired and stored in raw format on the server, which is redundantly backed up. Optionally, imaging data can be stored using a database such as OMERO.
Analysis: Data from the server is used as an input into an analysis workflow or pipeline. The code to run the analysis is version controlled using a remote server such as GitHub.
Communication: The analysis generates plots, which can be turned into figures and used for presentation. Analysis can easily be rerun if new data or new code is available or if requirements change. Plots, code, data, and database links are stored back in the ELN using the experimental accession number.
Key: Black arrows, experimental flow; gray dotted arrows, information stored in the ELN; black dashed arrows, working with data.

work exclusively on this topic, you only need one project. If you start a side project on transport of lysosomes, this could become a second project. Data sets then correspond to the experimental accession numbering system described above. Permissions can be set as required. For example, all lab members and the PI can belong to a group and permissions set in OMERO such that each lab member can only see their own data and the PI can see all data from everyone in the group.

You can think of your OMERO database as a rich extension of your lab notebook. For example, images can be individually annotated, tagged, rated, and commented on. An image can even have an attachment, which might be the experimental protocol or some analysis done on that image. Importantly, OMERO can link out to an entry in your ELN, and this is how it can integrate into your organization. It is possible to share data with collaborators using OMERO and also to upload data to public repositories such as the Image Data Resource (IDR), which runs using OMERO. There are other image database platforms available, but they tend to be tied to server or microscope hardware (e.g., ACQUIFER or ZEN image browser from Zeiss).

Sharing Your Data Externally

With internal data organization taken care of, you and others in your lab can easily access your experimental data. But what about sharing the data with the outside world?

In genomics and structural biology, there is a long history of data sharing over the Internet. For cell biology, data sharing has traditionally been restricted to certain large-scale projects. Publication requirements are changing, and many journals now request that raw data be made available at publication. Data repositories are available to help you make your data available to the world. For example, Image Data Resource from OpenMicroscopy is set up to host images for access over the Web.[4] Other repositories include Figshare, Dryad, and Zenodo. An alternative strategy is to host the data yourself and simply put it on the Internet. The main problem with this is that you are responsible for its permanence on the Web. Server infrastructure may need setting up, and it definitely will need ongoing maintenance. And after all that trouble, your valuable data resource might be difficult to find as a lone presence on the Web. Moreover, what happens if you move to another institute or retire?

Another scenario is that you want to share your data externally, just not with the whole world. Maybe you have a collaboration with a research group at another institution and need to give them access to your data. Image databases such as OMERO can be configured to allow external users to log in and access images. For very large data sets, however, transferring data via the Internet can be too time-consuming, even with a very fast connection. Couriering a hard disk containing a few terabytes of data is a low-tech but effective solution!

Backing Up Your Data

Data loss can be catastrophic, and you need to put in some effort to make sure it does not happen to you. "Save early and save often" applies to individual documents, which could get corrupted or lost with a software crash. However, the loss of one file that you have spent several hours working on is nothing compared to loss of an entire data set or even the total loss of all of your data. Your data management plan should incorporate a secure place to store your data. Maintaining the server hardware is unlikely to be your responsibility, but it is worth checking that the server on which you are storing your data is properly maintained. Is it backed up regularly? To a remote location? What is the disaster recovery plan? How long would it be before you could access your data again after a hardware failure or a fire?

An example setup is that data from a microscope session is copied directly to your share on a server in raw format. All shares on the server are backed up on a schedule to an identical server at a separate location. This schedule could be snapshots taken every hour for a day, every day for a month, every month for a year. This allows the user to recover old data in case of accidental deletion, file corruption, or some other error. Ideally, the mirror can be brought online quickly in the case in which the primary server goes down, so that everyone can keep working. Once you have a system in place, check that it is working (i.e., can you restore from the backup?).

Typically a lab member would use files by pulling a copy of the files down from the server and working on them locally. The raw files always remain on the server—untouched. Any analysis that happens locally can also be backed up on a schedule (e.g., using Time Machine for Apple computers, or third-party software such as Genie Timeline for Windows). Ultimately all analysis files and associated materials should be stored back on the server.

Cloud storage solutions became popular a few years ago, and they are great for small amounts of data and for working collaboratively on projects. The advantage is that you can access your files anywhere in the world and share them easily. You also do not have to worry about storage infrastructure or backup schedules; a company takes care of this for you. However with large, multi-terabyte data sets, for reasons of speed and cost, it is better to have files stored using hardware on your own campus. Find out about the services the IT department at your institute can offer.

◆ *Golden Rules*

- Being organized is helping the future you.
- Organize everything around experimental accession numbers.
- Be consistent with dates and naming files.
- The ELN is the center of your organization, interfacing with lab databases and protocols.
- Store the raw data and do not alter this in any way.
- Back up, back up, and check your backups.
- "Do not be that person" in the lab who is not organized.

CHAPTER 3

Imaging Data

I n this chapter, we will look at what information images contain and introduce the software that we can use to extract this information. In Chapter 4, we will look in detail at image analysis and processing.

SOFTWARE SELECTION

There are a range of software packages available to analyze data in an image analysis workflow. Examples include Fiji/ImageJ, Icy, CellProfiler, IMOD, Imaris, MATLAB, Python, R, LabVIEW, IGOR Pro, and more. Each have strengths and limitations—which one will you choose?

In cell biology, analysis workflows typically contain one or more image analysis steps, followed by some number crunching. Ideally, this is done as a *pipeline* (i.e., entirely within one software environment). In practice, it is often not possible. First, the software may work for one step of the workflow but not another. For example, it is possible to track cell movements in ImageJ, but to analyze those tracks and make publication-quality graphs you need a different software package. Second, even if it is possible to write an entire workflow within one package, it might take too long to do so. Someone else may have already written a procedure to do the analysis step you want to do and, chances are, it is not written in your favorite package. It is important to be flexible and to be open to using several different packages to build your workflows. Being loyal to one software package is admirable, but it can slow you down significantly.

Another consideration is software availability. Software licenses can be very expensive and you might not always have access to the software yourself. Open-source software is preferable because whatever code you write can be used by anybody, and this is, of course, more useful to science. Note that some closed software packages often have versions with limited functionality available for users without a license, so commercial software should not be dismissed out of hand.

A final thing to consider is the user base of the software. Getting help is an important part of coding, especially when you are learning (see Getting Help in Chapter 6). Most software packages have dedicated online forums or mailing lists where users can post questions and get advice. If there are few users, getting help may be difficult or slow. On the other hand, if there are many users, you may get lots of advice but of lower quality. If the goal is to share your software with others, writing for a widely used package means that it is more likely people will use your software.

For the purposes of this book, we use ImageJ for image analysis and R for number crunching. Currently, Fiji and RStudio are very useful implementations of ImageJ and R, respectively. Fiji is an installation of ImageJ that is preloaded with lots of useful plug-ins.[5,6] RStudio is an integrated development environment (IDE) for R that has many useful features.[7]

Fiji

Fiji (Fiji is just ImageJ) is a software bundle containing the image analysis program ImageJ together with lots of useful plug-ins and features.[5,6] The idea was to create a "batteries included" version of ImageJ that was ready to go for most users. ImageJ was first developed by Wayne Rasband at the NIH. Originally called NIH Image, it was ported to Java[*] so that it would work cross-platform. Now termed ImageJ 1.x, it is still maintained, but ImageJ2 (IJ2) is a rewritten and much extended version. IJ2 is more extensible and has many useful features, such as the ability for plug-in writers to deploy code via update sites[8] (see Sharing Your Code in Chapter 6). Fiji relies on IJ2 (yet also carries IJ1 for backward compatibility), and this is the software we will work with in this book.

When you open Fiji, the first window you'll see is the main window (Fig. 3.1). Besides image windows, other useful windows can be spawned using menu commands. The ROI (region of interest) manager, macro programming window, or image properties can all be invoked when needed. The results window, for example, pops up whenever a measurement is made (by pressing *command + m* [Mac] or *Ctrl + m* [PC/linux]). In Chapter 4 we will look at how outputs from Fiji can be saved as a csv file for importing into R to do further analysis.

RStudio

RStudio is an IDE for R, a programming language for statistical computing.[9] Both RStudio and R need to be downloaded and installed. A number of useful open-source packages and libraries are available from CRAN (Comprehensive R Archive Network; https://cran.r-project.org) or other sites on the Internet. It is useful to put

---

[*]Java is a general purpose computer programming language. Programs written in Java are designed to be run as stand-alone applications, with few dependencies.

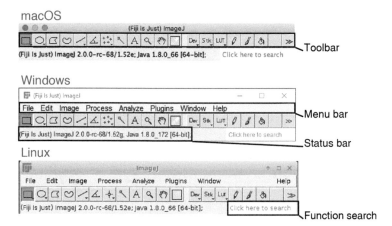

FIGURE 3.1. Fiji main window.
The main window of Fiji contains the menu bar, tools, status bar, and function search.

all R projects together in a folder in your home directory (e.g., ~/Documents/ RProjects/). To begin a new project, choose *File > New Project...*, select New Directory, New Project, and then give the new directory a name and designate ~/Documents/RProjects/ as the parent directory. RStudio will now initiate itself with this project (Fig. 3.2).

You can make a new R Script in the project using *File > New File > R Script* and use this to write some code to execute on the command line using the "Run" button (Fig. 3.2). When it is time to exit, RStudio asks if you want to keep the data in the workspace. It is best practice to click "No." To make this the default, click *Preferences > General* and set *Save workspace to .RData on exit* to "Never."

Keeping the data means that any errors get carried over to the next session. More importantly, for reproducible working, you should not rely on old data being there in RStudio when you restart it. As described earlier, you should have the raw data and the code to process it, and any outputs should be disposable.

WHAT IS AN IMAGE?

Images are just numbers, organized in a 2D matrix. Each pixel represents a number at that location in the matrix. Doing "image analysis" is simply making calculations using subsets of these numbers.

Figure 3.3 shows a mitotic cell expressing a GFP-tagged kinetochore protein (kinetochores are the complexes that attach chromosomes to the spindle during mitosis). The kinetochores appear as small bright spots. If we zoom in on one of these spots and then look at the values that correspond to the spot, we can see that the

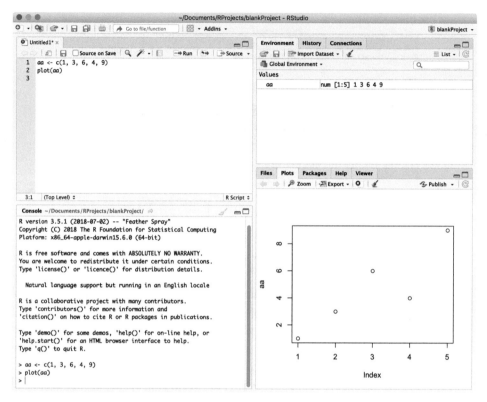

FIGURE 3.2. RStudio desktop IDE for macOS.
Typical layout of a project in RStudio. Writing code is done on the *left*. Support is on the *right*. R Scripts or R markdown can be composed in the *top left* window, whereas the console (*bottom left*) is a command line to manually enter commands. The Environment window (*top right*) shows objects created or a history of commands. *Bottom right* there is a file browser, which can also show help files, plots, and more. In the figure, a two-line script has been composed and run to generate the simple plot.

bright pixels have high values and the surrounding pixels correspond to lower numbers.

In mathematical jargon, an image is a two-dimensional function, $f(x,y)$, where x and y are the spatial coordinates of the image and the amplitude is a discrete value at that point.

The range of values in the image depends on the "depth" at which the image was captured. In the image above, the depth is 8-bit.[†] This means that the range of

---

[†] A bit is a "binary information digit" that encodes information in computing. A bit may be one of two values, 0 or 1. An octet is 8 bits, which can represent 256 discrete integers.

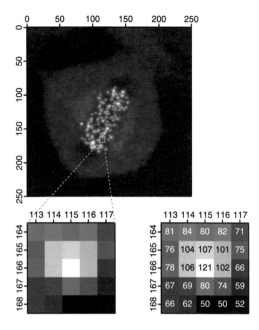

FIGURE 3.3. Images are just numbers.

possible values is 0–255. Depending on the detector in the camera used to take the image, a variety of bit depth and corresponding gray values are possible. The most common bit depths on modern cameras are 12-bit and 16-bit.

Bit depth	Basis	Gray levels	Integer range
8-bit	1×2^8	256	0–255
10-bit	1×2^{10}	1,024	0–1,023
12-bit	1×2^{12}	4,096	0–4,095
14-bit	1×2^{14}	16,384	0–16,383
16-bit	1×2^{16}	65,536	0–65,535

Publishers and most nonscientific software packages prefer 8-bit images. They cannot use or display higher bit depths. For image analysis, we use the images at the depth they were captured and only downsample to lower bit depths if it is necessary.

Note that the way that an image is displayed on a computer screen does not necessarily represent the underlying data. For example, the image above is displayed with black as the minimum value in the image and white as the maximum value, rather than black being 0 and white being 255. This is a type of image manipulation known as *contrast adjustment* (see Chapter 7 for further discussion of acceptable vs. unacceptable contrast adjustments).

Image Formats

The pixel values in an image are stored in a file and the standard format is tagged image file format (TIFF). TIFFs are lossless, support gray scale or color, and can be 8-bit or 16-bit. They are completely portable and can be used in any software or operating system.

Other less portable formats exist that are lossless and support full bit depth and multidimensional imaging—for example, custom formats from microscope companies: nd2, zvi, lsm, ids, etc. It is possible to open these images in certain software packages and do limited analysis. However, the approach advocated in this book is to bring all formats into Fiji and analyze them there, which means conversion to TIFF, the working format in ImageJ.

Note that there are a number of other image formats, which are popular for domestic photography or for display on the Internet—for example, JPEG, GIF. These formats are lossy[#] and they only support 8-bit; therefore, they should not be used for analysis.

Fiji uses a Java library[**] called Bio-Formats to read in custom image formats. Bio-Formats can read many different image file formats and understand their metadata, enabling images to be imported and displayed with the correct dimensions and scaling.

Because TIFFs can only be 8-bit or 16-bit, cameras that capture at intermediate bit depths (e.g., 12-bit) are saved as 16-bit TIFFs. Technically, the way this works is that either the upper bits are reserved or the image is scaled to 16-bit by counting in 16s.

Image Types

In Fiji, there are five image types:

- 8-bit—display 256 (0–255) gray levels (integers only)
- 16-bit—display 65,536 (0–65,535) gray levels (integers only)
- 32-bit—display >4 billion gray levels (real numbers) (pixels are floating point [not integers] and can have any intensity value including NaN [Not a Number][‡])

[#] Lossy compression is a way to make files smaller by discarding information and using approximations.

[**] A Java library is a collection of code written in the computing language Java. Libraries are used by different programs to perform routine tasks, such as image loading and writing.

[‡] This is useful when making ratio images (dividing one image by another). The resulting pixels will be real numbers and they might not be calculable. In the case of division by 0, this can be represented by NaN.

- 8-bit color—display of up to 256 colors encoded via a lookup table that is stored with the file (special case)
- RGB color—display of red, green, and blue colors, each encoded as 8-bit values—sometimes called 24-bit images (32-bit is possible for RGBA, with the fourth 8-bit encoding alpha [opacity])

Color in computing generally refers to the display of red, green, and blue values in RGB space. Here each pixel has three values, one each for red, green, and blue. This triplet determines the color via an *additive* color scheme. In 24-bit images, in which pure red and pure green overlap, the pixel value will be (255,255,0). This corresponds to 100% red channel, 100% green channel, and 0% blue channel and gives pure yellow for those pixels. Gray-scale images are actually displayed on an RGB monitor using equal values for each channel (i.e., a pixel value of 42 is displayed as 42,42,42). Print media uses *subtractive* color schemes, typically CMYK (cyan, magenta, yellow, and black). Here combinations of inks give new colors because of their absorbance and reflectance properties—for example, cyan and magenta give a blue color. Not all colors in RGB space are represented in CMYK, and as a result, microscopy images tend to look dull and less vivid after conversion. You will not need to use CMYK unless you are preparing an image for printing.

In Fiji, color is dealt with in three different ways. The first way is pseudocoloring. A gray-scale image is displayed with a continuous lookup table (LUT). Here each pixel has a single value, and those values are displayed using a color lookup table. The gray-scale value is "looked up" by the table, and the corresponding RGB color is found and displayed. The second way is RGB color in images. Here each pixel has three values, one each for red, green, and blue. This triplet describes the color to be displayed in RGB color space. Because each pixel has three values, RGB images can be split into three gray-scale images or indeed made by combining three gray-scale images. The third way is composite colors, in which the channels can be switched, turned on and off, and pseudocolored.

One final note on color. Two other color schemes besides RGB are available in ImageJ: HSV/HSB (hue, saturation, and brightness/value) and Lab color schemes (lightness, color(a) and color(b)). You are unlikely to need these color schemes, but they are mentioned here for completeness, because they are valid image types.

Multidimensional Image Files

As we saw, a single gray-scale image is a 2D matrix, in which the x and y directions correspond to the width and height of that image. Higher dimensions are possible, allowing for any or all of the following:

- color (*C*)—multiple *channels* can be acquired
- *z*-depth (*Z*)—*z*-series have multiple *slices* for each *x,y* location
- time (*T*)—time series have multiple *frames*, one for each point in time

To further complicate things, identical sets of images that have all of these higher dimensions can be captured at different spatial locations in a multipoint experiment and stored in one file! We will ignore this for now and focus on the images captured at a single location.

A single gray-scale image is the most simple image type ($C = Z = T = 1$). Once we have more than a single image, it is referred to as an image *stack*. This could be one multichannel image with, let's say, four channels ($C = 4$, $Z = T = 1$), or it could be a *z*-series of 10 confocal images with one channel ($C = 1$, $Z = 10$, $T = 1$),[##] or a movie of one channel at a constant *z*-depth over time at 1 frame/sec for 1 min ($C = Z = 1$, $T = 60$).

Image *hyperstacks* are a notch up in complexity. Multichannel *z*-series, multichannel movies, single-channel *z*-series over time, or even multichannel *z*-series over time are all examples of hyperstacks. Such images are sometimes referred to as a 5D image, because *X*, *Y*, *C*, *Z*, and *T* are all greater than 1. The convention is for images to be organized in *XYCZT* order, but there is no consistent terminology regarding dimensions—for example, a 4D image may refer to a multichannel *z*-series or a multichannel movie. The order refers to the position in which each individual image is stored within the file. In an *XYCZT* file composed of two channels, three *z*-slices, and 10 time points, the first image is channel 1 at the first *z*-slice and time-point, the second is channel 2 at the same position, the third is channel 1 again at the next *z*-position, still at the first time point and so on. The seventh image would show channel 1 at the first *z*-slice for the second time point, and there would be 60 images in total. These details matter if you would like to reorganize the hyperstack, creating substacks, or de-interleaving images.

More dimensions means larger files. Capturing an overnight experiment of two channels, seven *z*-slices at 3-min intervals, means a total of 3360 images. If the images were 1024 × 1024 (i.e., 1×10^6 pixels) at a depth of 16-bit, this works out at an approximate file size of 7.1 GB. Note that many programs will not open TIFFs larger than 4 GB, and opening large hyperstacks in Fiji is possible but ultimately depends on the RAM available. Using virtual stacks can get around these problems by displaying the active image "on the fly" as it is read from disk.

[##] When the *z*-dimension is added, what were pixels are now referred to as *voxels* (volumetric pixels).

Metadata

Metadata are data that give extra information about the image series or data set—for example, which dimensions of the image have been captured. Additional metadata, which can be specific to the file format, store information about the way that the image was captured—for example, what objective lens was used, what the laser power was, what gain setting, and so on. This information can be read when the file is imported. It is used (among other things) to display the image correctly.

You need to check that the microscope software is accurately recording the metadata. If your microscope has a manual nosepiece, you probably need to tell the software which objective lens is being used. If you don't, the downstream analysis might not be scaled correctly.

To examine the metadata after opening the image in Fiji, click *Image > Show Info...* or use command + I (Mac), Ctrl + I (PC). The window will show something like this:

```
HeLa_GFPRab5_mChTubulin_1 BitsPerPixel = 16
HeLa_GFPRab5_mChTubulin_1 DimensionOrder = XYCZT
HeLa_GFPRab5_mChTubulin_1 IsInterleaved = true
HeLa_GFPRab5_mChTubulin_1 IsRGB = false
HeLa_GFPRab5_mChTubulin_1 LittleEndian = true
HeLa_GFPRab5_mChTubulin_1 PixelType = uint16
HeLa_GFPRab5_mChTubulin_1 SizeC = 2
HeLa_GFPRab5_mChTubulin_1 SizeT = 253
HeLa_GFPRab5_mChTubulin_1 SizeX = 1024
HeLa_GFPRab5_mChTubulin_1 SizeY = 1344
HeLa_GFPRab5_mChTubulin_1 SizeZ = 1
HeLa_GFPRab5_mChTubulin_1 Camera/Detector = Rear camera (Hamamatsu
    C10600-10B ORCA-R2)
HeLa_GFPRab5_mChTubulin_1 Channel #1 = 488 nm for imaging
HeLa_GFPRab5_mChTubulin_1 Channel #2 = 561nm for imaging
HeLa_GFPRab5_mChTubulin_1 Name = HeLa_GFPRab5_mChTubulin_1
HeLa_GFPRab5_mChTubulin_1 Objective magnification = 100.0
HeLa_GFPRab5_mChTubulin_1 Pixel height (in microns) =
    0.06896478832687149
HeLa_GFPRab5_mChTubulin_1 Pixel width (in microns) =
    0.06896478832687149
HeLa_GFPRab5_mChTubulin_1 X Location = 0.0
HeLa_GFPRab5_mChTubulin_1 Y Location = 0.0
HeLa_GFPRab5_mChTubulin_1 Z Location = 0.0
HeLa_GFPRab5_mChTubulin_1 Z step (in microns) = 1.0
```

This tells us lots of useful information and allows Fiji to scale the image correctly (see Fig. 3.4). We know, for example, that there are two channels (because SizeC = 2), and we know that the fluorophores were excited with 488- and 561-nm light. The only remaining challenge is to figure out what was imaged in which channel. Some microscope software packages allow you to enter which fluorophores were imaged and save this as metadata. If not, the simple solution is to include this information in the filename, using the order of channel capture. In the example above we can see that GFP-Rab5 and mCherry-Tubulin were imaged in HeLa cells. Naming files in this way is simpler than relying on a table of conditions in your lab notebook, which requires cross-referencing and is prone to error. If you are using OMERO to organize your images, you can tag and label the images in lots of different ways to ensure you can easily find your images and work out what exactly was being imaged.

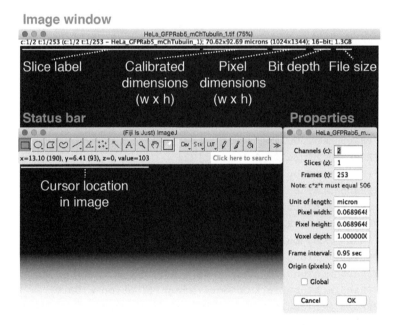

FIGURE 3.4. Properties of an image in Fiji.
The properties of an image are derived from its metadata. They allow the image to be displayed correctly. The Infobar at the top of the image window shows the dimensions of the image, calibrated according to the metadata. If the pixels are spatially calibrated, the cursor location will be displayed in the status bar in pixels and in calibrated units, along with the intensity of that location. The dimensions of the image can also be viewed using *Image > Properties*. If there is an error and the image needs to be rescaled, you can do so using *Analyze > Set Scale… .*

Image Transformation

Images can be transformed using either positional or intensity transformations. Positional transformations include rotation, transposition, skew, translation, and scaling. They relocate pixels or resample the pixels in their new location. In intensity scaling, all the pixels remain in place but their values are altered (Fig. 3.5).

Downsampling is the simplest form of intensity scaling—for example, the creation of 8-bit images from files that were captured at a bit depth greater than 8-bit. A 12-bit image contains pixels on a scale from 0 to 4095. To convert to 8-bit, we simply assign new values to the original pixels. Thus, 0–15 become 0, 16–31 become 1, and so on. You can see immediately that we are losing information in this process. It is for this reason that it is best to analyze images at the bit depth at which they were originally captured. But there are some good reasons for downsampling. Publishers require 8-bit images and computer screens are 8-bit, so 8-bit data are required to display the information correctly. Sometimes, image processing steps will only work on 8-bit images, and so downsampling is required to do these steps. Although it is possible to upsample an image from 8-bit to 16-bit, there is no gain in information in doing so. Changing the image type in this way in Fiji is done using *Image > Type*. As an example, if we start with a 16-bit image that has minimum and maximum values of 1939 and 43,528, after changing the image type to 8-bit, these values become 8 and 170. By default, ImageJ converts 16-bit (and 32-bit) images to 8-bit by linearly scaling from min–max to 0–255, in which min and max are the two values in the "Display range" (*Image > Show Info*). If the conversion is done on a stack, the whole stack is scaled according to the min–max values of the currently displayed slice. This behavior can be disabled by unchecking "Scale When Converting" in *Edit > Options > Conversions*.

Besides conversions, there are other intensity scaling methods to change the display of the pixels or the values of the pixels themselves (Fig. 3.5). You can play along in Fiji using *Image > Adjust > Brightness/Contrast* or using Shift + Command + C (Mac) or Shift + Ctrl + C (PC). Any adjustments made in Brightness/Contrast only change the way that the image is displayed, and this is demonstrated by the line plot shown in Figure 3.5. Analysis performed on the image before or after adjustment will give identical results. To commit these changes to the image and actually change the pixel values, click *Apply*. Now any analysis will be different because the pixel values have been modified. We normally only want to change the display of the image to help with analysis of the original values, so pressing Apply should only be done if you are sure you want to change the pixel values.

A simple transformation is to invert the image (using *Edit > Invert*). Here 0 in the input image is displayed as 255 in the output and 255 as 0. Note that LUTs control

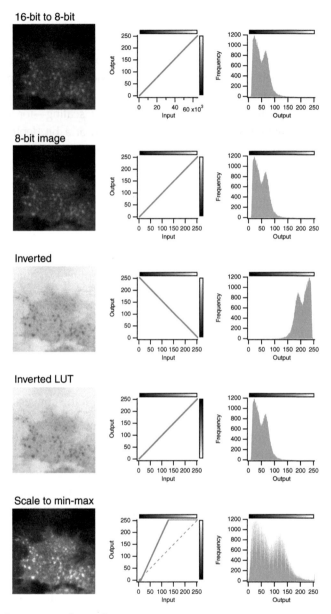

FIGURE 3.5. Image transformations.
Different image transformations are shown as an image (*left*) and a representation of the input data and output (*middle*) that represents the transformation and how it changes the way that the image is displayed. If the transformation is applied, the pixel values are changed. The result changes the image histogram (*right*), which shows the frequency of pixels in the image with the given value. The image shows the footprint of a cell expressing a GFP-tagged protein found in vesicles, as imaged by total internal reflection fluorescence microscopy.

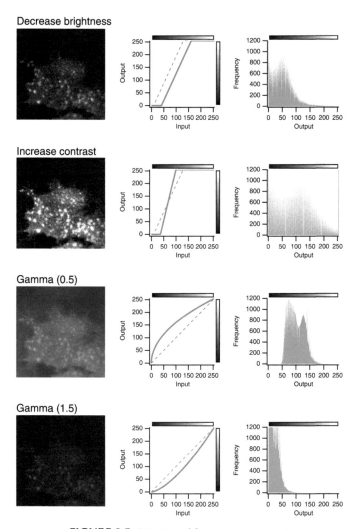

FIGURE 3.5. (*Continued from previous page.*)

how the image is displayed, and the same effect can be achieved by inverting the LUT rather than doing any image transformation. To scale to the minimum and maximum pixel values, we can enter them by using the "Set" option. Clicking "Auto" finds two values inside the minimum and maximum pixel values that will saturate a very small number of pixels (here they are set to 9 and 129). Note how the slope of the line has changed. This is referred to as "contrast stretching." Now any pixels above 129 will display as white (255), regardless of their value. Conversely pixels below 9 will be black (0). This is known as clipping, trimming, or saturation. Any detail in the clipped region will be lost. As this manipulation effectively discards information, it should be done with caution (see Chapter 7 for a discussion of image manipulation). In the Brightness and Contrast tool in Fiji, brightness can be altered, which moves the line to the left or right, whereas contrast changes the slope of the line.

All of the transformations described above are linear (i.e., the line representing input and output is straight and not curved). Nonlinear transformations are possible, the most common type is γ (*Process > Math > Gamma*). For linear transformations, $\gamma = 1$. If $\gamma < 1$, it makes the image brighter overall, but it boosts dim features more than brighter ones. The converse is true with $\gamma > 1$ (Fig. 3.5). All nonlinear transformations must be disclosed when presenting your data.

Intensity transformations can help visualize your imaging data, but they can only do so much. Taking the best-quality images is the key to successful image analysis.

The bottom line here is do not do any positional transformations of your images. All images should be analyzed as captured. If images are scaled, skewed, or rotated arbitrarily, the pixels are resampled to assemble the new image. This means that the original data have been changed and are no longer fit for analysis. These manipulations can be difficult to reproduce, so they are simply best avoided altogether. An exception here are transformations that do not affect the data: transposition or translation in which the image is rotated by ±90° or flipped horizontally or vertically. An extension to this advice is that if you are scanning a gel for analysis, you should scan it so that the bands and lanes are horizontally and vertically aligned. Do not rely on the software to correct it afterward.

IMAGING INFORMATION

So far, we have discussed imaging *data*, but what about imaging *information*? What *information* is contained in the images you want to analyze? Taking pictures of cells is easy; capturing images of cells that contain useful information is hard.

In any imaging experiment, we make many assumptions about what we are imaging and what the features in the image represent. For example, let us consider a two-color immunofluorescence colocalization experiment. The analysis assumes that each channel of the image contains information from one fluorophore only, and that this accurately reports the protein or feature that we are interested in imaging. So, care must be taken to set up the microscope and capture images in such a way that there is no bleed-through of fluorescence from one channel into another. How well are the channels registered with respect to one another? In some microscopes, differences in the lightpath for each channel can mean that pixel locations may be offset or distorted. And what about the sample: Are the antibodies specific for each protein that you are studying? Is there any cross-reactivity between the secondary antibodies? If you have added a fluorescent tag to the protein you are studying, has this interfered with its function or its localization?

It is worth taking time to ensure that you are capturing high-quality imaging *information* rather than just capturing imaging data. There is a saying in data analysis—GIGO—meaning *garbage in, garbage out*. This means that if your starting data are poor quality, it does not matter how good your analysis methods are, the end result will be meaningless. If the microscope is set up badly, or if some other aspect of the experiment is suboptimal, it might not be possible to get any meaningful information out.

Trade-Offs in Imaging

Everything in imaging is a compromise. Ideally, we want lovely bright samples with strong signals. A simple way to achieve this is high-density labeling. However, we do not want to overexpress our protein (tagged with a fluorescent protein) and would prefer to image it at endogenous levels. So we could compensate for lower labeling by increasing the illumination. Now we have a new problem: Our sample bleaches quickly. So we increase the gain on the camera, but now we have more noise. And so it goes on.

There are four desirable criteria in imaging, and enhancing any one of these compromises the other three:

1. High temporal resolution
2. Low photobleaching and phototoxicity
3. High spatial resolution
4. Multiple channels, multiple sections

For example, to get high temporal resolution, you need to take pictures more frequently, which means more photobleaching and phototoxicity. If you want to image really fast, it might mean compromising spatial resolution by binning the images. When imaging, your mission is to optimize the parameters that are important for your experiment. This involves making trade-offs in the imaging conditions until you can reliably get good-quality images containing useful information.

Focus and Dealing with Drift

The cells and structures within the cells need to stay in focus throughout a movie. If there is drift in the z-direction, this can prevent images from being analyzed. If the microscope is well-maintained and is on a stable, vibration-isolated table, then the most common cause of z-drift is thermal changes in the room. Using an enclosed chamber, which is preheated and at a stable temperature, as well as having good temperature control in the room, will minimize the impact of these fluctuations. Drift in the *X–Y* direction is easier to correct by reregistering images afterward, but it is still preferable to have no movement beyond the biological movements happening in the sample.

Phototoxicity and Photobleaching

Phototoxicity and photobleaching occur in response to illumination used for imaging. Photobeaching is the reduction in fluorescence emitted from the sample during repeated imaging. Phototoxicity is a phenomenon in which cells become damaged and ultimately die as a result of illumination. Photobleaching is specific to the fluorophore, but phototoxicity can happen at many wavelengths of light. A small amount of photobleaching can be corrected computationally (using *Image > Adjust > Bleach Correction*), but more severe bleaching needs to be minimized when the images are captured. It is important to assess toxicity by checking if the cells you are imaging show signs of stress (blebbing, reduced motility, etc.). Minimizing light exposure, either the time used for capture or the intensity of illumination, is the best way to offset this damage. Another strategy is to use a fluorophore with a greater quantum yield to improve the signal obtained per unit of excitation light.

Choice of Fluorophores

Understanding the setup of the microscope you are using is essential. Which filter sets are on the microscope? What wavelengths are available for excitation? A typical setup on a widefield microscope is a filter cube for each channel, with a narrow band of excitation light, a dichroic mirror, and a bandpass emission filter to collect the fluorescence. On a point-scanning confocal microscope, laser lines allow more precise excitation—for example, 488-nm or 561-nm light. Emission is usually collected through filters that can be tuned in order to optimize collection. Once you know what options you have for excitation and collection of emission, you can plan how to take the best images. Aim for no cross talk between channels. Cross talk occurs when the fluorescence from one fluorophore bleeds through into another channel (Fig. 3.6). There are interactive tools available on the Web that allow you to check how close the excitation and emission spectra are for the fluorophores you want to use, how efficient the excitation will be, how much light you will collect, and how much will be excluded. The properties of fluorophores are also found online. Some are brighter than others; some bleach faster. Select fluorophores that are well-matched to your microscope setup. You should aim to maximally

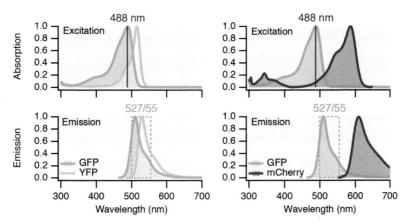

FIGURE 3.6. Cross talk between fluorescent proteins.
There is overlap between the fluorescence spectra of GFP and YFP (*left*), but less so between GFP and mCherry (*right*). Overlap is the area under both curves where the spectra coincide. Excitation with a 488-nm laser excites GFP, but also YFP, because of this overlap. In the case of GFP and YFP, the emission spectra also overlap, so collecting light with a 55-nm band-pass filter centered on 527 means collecting the GFP fluorescence but also the YFP fluorescence that was accidentally excited with the 488-nm laser. With mCherry, there is much less excitation of mCherry by 488-nm light. Emission of mCherry is essentially 0 using a 527/55 filter, so there is no measurable cross talk.

excite the fluorophore with the illumination light. Any loss of efficiency here means putting in more light or exposing for longer, which both result in more photo-toxicity or photobleaching.

Dynamic Range

For quantification it is important to capture images using identical acquisition settings. This means the same magnification, binning, illumination, filter sets, exposure time, gain, offset, and so on. Images can only be directly compared if the images are captured in exactly the same way. Switch off any "auto" features in the microscope software and give yourself full control of image capture.

Establishing the acquisition settings needs effort to get right. Image your positive and negative controls first and check that your acquisition settings will capture the full range of samples in your experiment (i.e., take an image of the dimmest cells you expect to analyze and also take images of the brightest). Aim to use the full dynamic range of the camera. Have a look at the histogram of the image or display the image using a lookup table that reveals saturated pixels. You need to avoid clipping of the brightest signals in the image while setting acquisition so that you see the detail of what you want to measure. Conversely, if everything is too dim, the range of values that you are measuring will be too restricted.

◆ *Golden Rules*

- Sample
    - Are all the reagents and fluorophores well-validated?
    - Is the labeling optimized?
    - Are the cells happy or well-mounted if they are fixed?
- Microscope
    - Is it set up right for your experiment (correct objective, excitation, and emission, etc.)?
    - Is the image saturated? Too dim?
    - Keep the settings the same for all samples.
    - Is the cell centered? Make sure that no interesting features are too close to the edge of the image.
- Images
    - Capture the raw data with no compression or transformation.
    - Do you have publication-quality images for every experimental condition?
    - Does the result make sense?
    - Remember GIGO (garbage in, garbage out).

CHAPTER 4

Image Processing and Analysis

We will now look at how to retrieve information from cell biological images. The tutorials will guide you through some common approaches in image analysis. You can use them in your work or use them as the starting point for your own analysis needs.

HOW TO ANALYZE AN IMAGE

Measuring what is happening in the whole image is rarely informative. Instead, we need to define regions of interest (ROIs) and measure what is happening in them. In the simplest case, we can make a rectangular selection and use *Analyze > Measure…* (or press "m") to retrieve the measurements. Now we are measuring just a subset of the pixels in the image. Figure 4.1 shows the difference between measuring an ROI and measuring the whole image.

You can experiment with this yourself to understand what is happening. Select an ROI in an image of your own, use the Measure command, now move the ROI to a darker part of the image, and Measure again. What happens to the Mean measurement when you do this? Try to predict the answer to the following questions: What would happen if the ROI stays in place, but the location of all pixels is randomized? What would happen if the location of the pixels *within* the ROI were randomized? As shown in Figure 4.1, randomization of pixel locations within the ROI makes no difference to the measurement, but randomization of all pixels causes a change. Location matters, and making the right selection is key.

We can make multiple ROIs and store them in the ROI manager. They do not have to be the same type: We can have rectangular, freehand, or other ROI types. The type of ROI depends on your scientific question and what you are trying to do. Using *More > Multi Measure*, we can run the "Measure" command for all of the ROIs at once. If there are multiple frames, we get information for each ROI for each frame. The ROIs from the ROI Manager can be saved so that they can be reused for reproducibility.

Any measurements get directed by default to the Results window (Fig. 4.1). Which parameters are shown in the Results window are customizable. Using *Analyze > Set Measurements…* allows the user to specify which parameters (displayed as columns) are displayed in the Results window, following the Measure

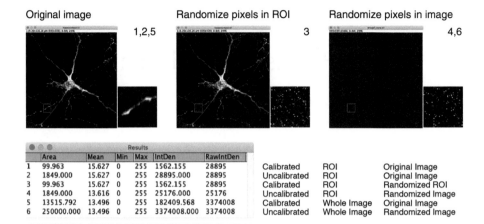

FIGURE 4.1. Getting information from images with the Measure command.
A confocal image of a neuron has been imported into Fiji. We can measure using a square ROI over a couple of synapses, or we could analyze the whole image. Notice the effect of randomizing the location of pixels within the ROI, and compare this with the effect of randomizing the location of pixels in the whole image. The position of the ROI is the same in all cases.

command. The *Area* of the ROI is shown. These are real units provided the image is correctly calibrated (99.963 μm²). If the image is uncalibrated, the result is the number of pixels in the ROI (1849 = 43 × 43-pixel ROI). The *Mean* (sometimes called the mean pixel density) is the sum of all pixel values in the ROI—also known as the raw integrated density (*RawIntDen*)—divided by the total number of pixels. The *IntDen* is the Area × Mean. In an uncalibrated image, RawIntDen and IntDen are equal (see measurement 2 in Fig. 4.1). You can calculate the number of pixels in the ROI of a calibrated image by multiplying the Area by the pixel size of the image (if you know it), or by dividing RawIntDen by the Mean. Knowing how these values relate to each other is useful when doing analysis. Using *Analyze > Set Measurements…*, other useful measurements can be included in the Results window. For example, *Min* and *Max* values show the minimum and maximum pixel values in the ROI. Including *Bounding Rectangle* is useful to record the ROI position in the same file as the measurement you make.

▶ Tutorial: Quantifying Cell Protein Levels from Immunofluorescence Images

Some cells have been stained with an antibody to a protein that we want to quantify. We have normal cells (condition A), cells that have the gene encoding our protein knocked out (condition B), and our test condition that we suspect reduces protein levels (C).

Here, we will manually measure the fluorescence intensity of a cell. We need to correct for background, which may vary from image to image. The resulting measurement is sometimes referred to as CTCF (corrected total cell fluorescence).

1. Open the image in Fiji.

2. Make your selection using the appropriate tool (see Note 1).

3. Add the ROI to the ROI Manager (press *t* or select *Edit > Selection > Add to Manager*).

4. Now draw a rectangular ROI away from the cell to measure background, and add that to the ROI manager.

5. Select *More > Multi Measure* to measure both ROIs.

6. Save ROIs using *More > Save*.

Note 1: We will draw around the cell body manually, to measure total cellular fluorescence. For other applications you may want to draw a small ROI in the cytoplasm or draw around an organelle.

The value from the second ROI (background) is subtracted from the measurement from the first ROI to give a measure of total cellular fluorescence (Fig. 4.2). This is a simple, manual example in which the measurements need to be repeated for all cells in the image and for many images, before being averaged to get a measure for protein levels under this experimental condition. Because this is a manual analysis, the analyst should be blind to the experimental conditions (see Tutorial: Blinding Files for Manual Image Analysis in Chapter 6).

This tutorial demonstrates the general approach for this type of experiment, but if the number of images is large, such a manual approach can become tiresome, not to mention error-prone. Being able to automatically detect regions of interest would dramatically speed up this procedure.

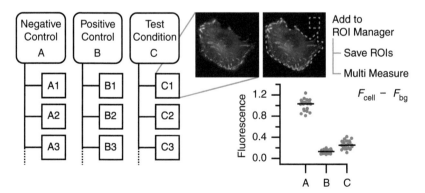

FIGURE 4.2. Manual analysis of cell protein levels.
Analysis of the experiment described in the text using a manual method to place the ROIs, save them, and extract measurements. The fluorescence of the cell is "background-subtracted" before plotting.

Segmentation

Segmentation is the process of dividing an image into segments for subsequent analysis. It can be done manually or automatically. Very often, the first step to automating image analysis is to dispense with the manual selection of ROIs and segment the image automatically.

The simplest form of segmentation relies on *thresholding* to identify objects in the image to analyze. Using thresholding, pixels with values in a defined range can be selected and then a binary image can be generated from this selection. A binary image means that pixels no longer have values on an 8-bit or 16-bit scale —they are now in either one of two states: false or true (normally encoded as 0 or 255 with no other values allowed). In a fluorescence microscopy image, we usually want to identify the regions with the brightest pixels and set those as true, with everything else being false. Selection of pixels above a given value can be done using *Image > Adjust > Threshold*. Values can be set using the sliders, and clicking "Apply" will create a mask based on these values.

Thresholding can be done automatically using one of several algorithms available in Fiji—for example, Huang or Otsu. This is preferable because the algorithms work *reproducibly* and this means that the routine can be scripted. Later, we will see how to automatically threshold a directory of images (see Chapter 6). However, determining which automatic thresholding algorithm to use is difficult. Fiji can show the result of all of the available algorithms to get you started. Select *Image > Adjust > Auto Threshold*, and then use "Try All."

Once a binary image is created, what can we do with it? We can use the image directly to count objects or measure their size. Alternatively, we can use regions defined by the mask to analyze features in the original image.

Using connected component analysis (CCA), a computer can group regions of "true" pixels into a distinct object. In Fiji, *Analyze Particles* is a CCA routine that can be used to identify objects in an image so that they can be counted, measured, and so on. Very small and/or very large objects can be filtered out using the dialog box. This is handy because sometimes one or two pixels can be above threshold and we do not want them to be treated as objects. Another common problem is that CCA connects two clearly distinct objects. Using a watershed routine is a way to separate these objects before running *Analyze Particles* (see Fig. 4.3).

Ticking *Display Results* and *Summarize* in the Analyze Particles dialog box provides analysis of the image on a per particle basis and as a summary. The columns in the results and summary windows are determined by what is selected in *Set Measurements...*. As a default, measurements are directed toward the mask image itself. Although this is useful, values for Mean Pixel Density and so on are simply related to the size of the particles and not to the intensity of the original image. Using the

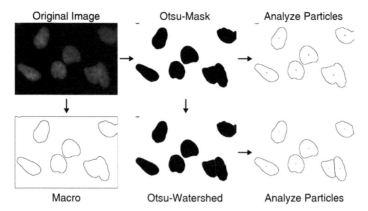

FIGURE 4.3. Segmentation of nuclei.
Original fluorescence microscopy image shows several nuclei stained with DAPI. Thresholding using the Otsu algorithm gives a binary mask. Note that Analyze Particles cannot distinguish between nuclei that are very close. Using Watershed after the mask creation can distinguish close nuclei. Finally, running the code example allows us to segment the perimeter of the nuclei.

redirect drop-down in the *Set Measurements…* dialog, it is possible to analyze the intensity of features in the original image. An alternative is to transfer the particles to the ROI manager and measure the ROIs in the original image.

The combination of automated thresholding and measuring using Analyze Particles is a way to collect data from many objects in an image for all images in a directory. This means that, provided the objects can be segmented accurately, the previous manual example can now be automated.

Further Segmentation Approaches

The process of thresholding and subsequent manipulation of the mask can allow for more sophisticated image segmentation. Besides watershed, it is possible to dilate or erode the mask to grow or shrink the objects, and other manipulations are possible. For example, this example macro will manipulate the nuclei mask in order to pick out the nuclear membrane (see Fig. 4.3):

```
1   setAutoThreshold("Otsu");
2   setOption("BlackBackground", false);
3   run("Convert to Mask");
4   run("Open");
5   run("Close-");
6   run("Outline");
7   run("Dilate");
```

The methods in this section are just the beginning. There are more advanced ways to segment images, such as local thresholding or by using *k*-means clustering[*] to define objects. In addition, preprocessing the image before segmentation and then using the mask to analyze the original image opens up yet more possibilities. How an image is segmented is probably the biggest challenge in image analysis. The routines that you develop will depend on your images and your scientific question.

Image Filters

As we saw in Chapter 3, images are simply numerical matrices. This means they can be manipulated using matrix math operations. Trivial examples are a matrix transpose to rotate the image 90° or subtraction of a constant to offset the intensity of the image. Using matrix convolution, images can be spatially filtered in sophisticated ways, which can aid detection of interesting features. Examples are smoothing and blurring to remove noise and thereby improve segmentation. Filters can be activated in Fiji using the menu *Process > Filters*. Choose *Convolve* to specify bespoke filters (some examples are shown below).

Matrix convolution is an operation that makes a new, filtered image $g(x,y)$ from the original image $f(x,y)$ using a coefficient matrix, called a kernel (ω). The kernel is usually much smaller than the image, 3 × 3 or 5 × 5 pixels. Mathematically the operation is defined as $f(x,y) * \omega = g(x,y)$.

The following kernel gives the same filtered image as the original:

1	0 0 0
2	0 1 0
3	0 0 0

Whereas with a flat coefficient matrix, we get a mean filtering effect known as smoothing. Shown below is a 3 × 3 kernel where all values are 1:

1	1 1 1
2	1 1 1
3	1 1 1

Working on each pixel in the image, the pixels in the 3 × 3 neighborhood are multiplied by the corresponding number in the coefficient matrix. The values are summed and then divided by the size of the coefficient matrix (Fig. 4.4). This is sometimes

---

[*] *k*-means clustering is a mathematical approach to classify *n* observations into *k* clusters. It can be applied to images to define similar pixels (observations) into distinct objects (clusters).

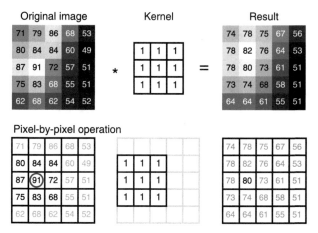

FIGURE 4.4. How image convolution works.
An original image is convolved using a "flat" 3 × 3 kernel. The result is a smoothed image. Convolution works pixel-by-pixel. The new value of the target pixel is calculated from the pixels in the neighborhood defined by the kernel, which is centered on the target pixel. Each pixel in the neighborhood is multiplied by the corresponding value in the kernel. These values are summed and then divided by the total value of the kernel. In the example, the target pixel (circled) is replaced by 80 because $(80 + 84 + 84 + 87 + 91 + 72 + 75 + 83 + 68)/9 = 80.4 \approx 80$.

called box filtering. Try it on an image and notice the effect using the *Preview* button. What happens if the filter is changed to a 5 × 5 or a 9 × 9 matrix of 1s?

In Figure 4.5, we can see some other uses of image filters. Using an approximation of a 2D Gaussian function as the kernel gives a blurring effect after convolution. Notice how the central value is the largest, and that the surrounding values decay symmetrically (Fig. 4.5). Smoothing and blurring are excellent methods for reducing noise in images, especially as a preprocessing step for segmentation.

To sharpen an image and highlight edges (regions of contrast), a simple kernel of negative values around a central positive value can be used (Fig. 4.5). Note that there are other kernels and approaches to achieve sharpening and edge detection. How could we change the kernel shown above to emphasize edges in a particular direction? Edit the kernel and toggle the preview switch to experiment with edge detection.

The filtering methods described so far are linear. There are also nonlinear filtering methods; in fact, thresholding is one such method. Median filtering (*Process > Filters > Median*) is a nonlinear method that reduces noise in much the same way as mean filtering, but it performs better with very spiky noise in the image. This sort of noise, sometimes called salt-and-pepper, consists of a very few, bright

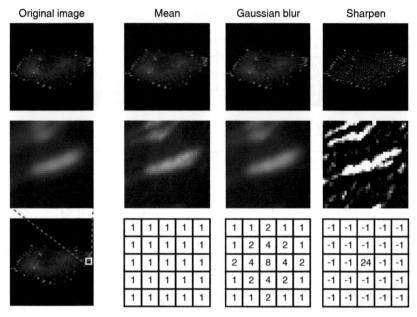

FIGURE 4.5. Examples of some common image filtering methods.
An original image is processed in three different ways: Mean, Gaussian blur, and Sharpen. The corresponding kernel is shown for each convolution; see text for details. The image is of focal adhesions in a mammalian cell; the zoomed region allows us to see more clearly the effect of the filtering methods.

pixels across the image. With a median filter, they get erased, because the median is insensitive to outliers, whereas with a mean filter, they remain as blurred spots (Fig. 4.6).

Gel Densitometry

Image analysis is not just for microscopy images. Gel densitometry can be used to quantify results from experiments such as western blotting. This method measures the density of bands on gels. It is only semiquantitative. In western blotting, proteins in a sample (perhaps a cell lysate) are separated by their size on a gel and then transferred to a membrane. This membrane is then probed for a specific protein using an antibody. The antibody is then detected either directly or indirectly using an amplification system such a chemiluminescence, before being imaged on film or with a camera. Films can be scanned and images analyzed by densitometry. There are variations on this method that claim to be "quantitative," but ultimately western blotting is a "semiquantitative" technique. Relative changes in protein levels can be

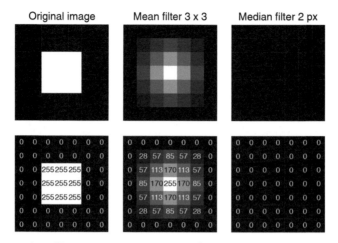

FIGURE 4.6. Median filter removes spiky noise from images.
Median filtering is a nonlinear method useful for removing noise from images. On the *left*, the original image has a bright white spot on a black background. Convolution with a 3 × 3 mean filter blurs the spot, whereas a median filter (width = 2 pixels) removes it completely. Images are shown *above*, and the same versions with their pixel values are shown *below*.

detected on a gel, and these can indeed be measured, but absolute quantification is very difficult to achieve. Reasons for this include:

- Detection methods vary in their sensitivity, linearity, and range of detection. Enhanced chemiluminescence and film together have a narrow dynamic range, notably at high protein levels.
- There is variability from gel to gel.
- Transfer can be uneven across the membrane.
- Standards are not routinely loaded to be able to do quantification.
- Equal protein loading is difficult.
- There is variation in the sample. A knockout clonal cell line will give a clear result, but in experiments in which a transfection is performed, the western result can be confounded by transfection efficiency.

With all of this negativity, is it worth even trying to quantify a western blot? The problem is that you need to show that your result is reproducible (i.e., can be seen clearly across several experiments). Space is often a problem in a paper, so showing every blot you have performed is not a solution. The best solution is to quantify the results, summarize them in a plot, and display the results from one experiment.

▶ **Tutorial: Quantifying Bands on a Gel**

Gels and films can be scanned in transmission mode at high resolution (600 dpi or higher) and loaded as TIFF files into Fiji.

To analyze the band density, use the manual method (Fig. 4.7A):

1. Draw a rectangular ROI around the band of interest. This box should be big enough to encompass the whole band and not stray into other lanes. Move it over all the lanes you want to quantify, and check that this one box size will work for all bands. Adjust if necessary.

2. Position the box over each band and save the position to the ROI Manager.

3. Now move the box to an area of background above or below the band of interest. Do this for each band, saving the positions to the ROI manager.

4. If the image shows dark bands on a light background, invert the image so that the bands have a higher pixel value than background.

5. Select all ROIs in the ROI Manager and click "Measure" (make sure that "Mean" is selected in *Set Measurements...*).

6. Save Results and the ROI set.

7. Load the results into R, and subtract the background from the respective band intensity.

8. Repeat for all blots, and plot the data.

Alternatively, use the built-in ImageJ method (Fig. 4.7B):

1. Draw a tall rectangular ROI around the first band of interest. The ROI should be wide enough to encompass the whole band and about twice as tall. Adjust if necessary; press "1."

2. Move the box to the next band and press "2."

3. Repeat for all bands, pressing "2" each time. ImageJ will annotate the boxes with consecutive numbers.

4. For this method, it does not matter if the bands are dark on a light background or vice versa.

5. Press "3" when you are happy with the placement of all boxes. You will see a new window with the vertical intensity profile of each band.

6. For each band, use the line tool to draw across the background area, as shown. Make sure the line crosses the band profile.

7. Now, using the Magic Wand tool, click inside the band profile, and repeat for each band.

8. The results shown are an area and can be saved and imported into R for plotting.

9. Repeat this process for all experimental repeats. Note that the gels need to be scanned with identical settings for this method to produce comparable results.

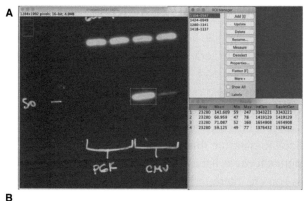

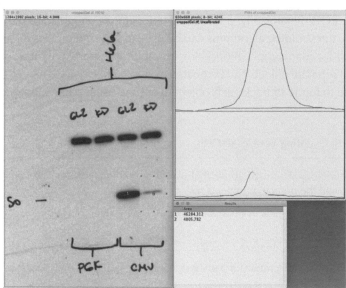

FIGURE 4.7. Gel densitometry.
There are two methods to quantify bands using Fiji. (*A*) Manual method. The image of the blot is inverted and ROIs over the band and the corresponding background area are measured. (*B*) ImageJ method. Vertical profiles of each band are generated using a key sequence and the area under the curve is measured.

An extension to this method is to also measure a loading control and normalize the intensity of the band of interest to the level of the loading control. A loading control is a "housekeeping" protein whose expression is supposedly constant. The problem with this is that the expression of many commonly used loading controls are themselves variable, and in any event their detection is subject to the same quantification problems as the target protein.

HOW TO ANALYZE A MOVIE

A movie is a series of images in a stack, and so the principles described above can be applied to each frame of a stack, in series. In a simple case, one or more static ROIs can be added to the ROI Manager and using *More… > Multi Measure*, the fluorescence for each ROI can be measured over time. The results from each frame are in rows in the results window, and the columns show measurements from each ROI. From this matrix we can plot fluorescence intensity over time, but other parameters, such as size, are difficult to measure this way.

This static ROI approach works if the duration of the movie is short, the frame rate is high, or the process you are interested in is not very fast. However, if the movie is long or the process is rapid, it is likely that static ROIs will fail. We need to be able to determine ROIs dynamically, and automated methods are great for doing this.

Again, thresholding algorithms can be used to segment movies for analysis. This works well for many applications—for example, if there is only one object in the movie or if separate objects can be treated as one. Otherwise we need to link objects from frame to frame to track each object over time using particle tracking.

▶ **Tutorial: Counting Vesicles in Cells**

The methodology described above that uses thresholding and Analyze Particles can be used to count the number of objects over time:

1. Open the image stack in Fiji and use a thresholding algorithm to binarize the image to reveal vesicles.

2. Select *Analyze > Analyze Particles…* .

3. Specify exclusion criteria for particles—for example, 9 pixel units for the smallest size.

4. Check the "Display Results," "Clear Results," and "Summarize" boxes, and click "OK."

The number of detected particles per frame is displayed in the "Summary" window. The "Count" column shows the number of particles detected per frame. In the "Results" window a measurement of each particle is provided. Again the contents of the "Results" window depend on the checked items in Set Measurements… . From these data it is possible to calculate a number of other useful statistics, such as the total vesicle area per frame, average vesicle size over time, and so on.

Particle Tracking

Particle tracking refers to the detection and tracking (usually over time) of individual objects, referred to as particles. The particles may be large (e.g., cells, nuclei), small (e.g., organelles), or subresolution spots of fluorescence (e.g., small vesicles below the resolution limit of the microscope).

Many routines for detecting and tracking particles have been developed and more will become available. Some routines will track in 2D only, whereas others will track in either 2D or 3D. Performance depends on the type of particle being tracked.[10] It is worthwhile to test a few routines to see what works best for your data. Notable examples include:

Software	Description	Reference
Fiji plug-ins	TrackMate, Manual Tracking, MTrack2, and ToAST	—
ComDet	ImageJ plug-in for particle detection and colocalization analysis	van Riel et al.[11]
u-Track	MATLAB routines for detection and tracking	Applegate et al.[12]
cmeAnalysis	Derived from u-Track; analyzes clathrin-coated pit dynamics	Aguet et al.[13]
KiT	Kinetochore tracking software for MATLAB derived from u-Track	Olziersky et al.[14]
CellTracker	ImageJ plug-in to detect cell movements	Shen et al.[15]
CellProfiler	Multiplatform software for image analysis	Carpenter et al.[16]
Icy	Image analysis package with spot and cell tracking capability	Chaumont et al.[17]
ImarisTrack	Commercial software Imaris from Bitplane	—

All routines work on a similar two-step basis. First, all particles must be detected in every frame with good reliability against the background. This is often done using a Gaussian-based method with some prefiltering of the image. Second, the particles must be linked to one another so that we can track them. This means that each particle in one frame must be connected to each particle in the following frame, with the possibility that some particles may disappear and that new particles may appear. This is done by applying a cost to each possible linkage between particles. The most straightforward is distance, such that the nearest particle in the following frame is most likely to be the same particle; although other parameters such as shape or intensity can be considered. The lowest cost linkages are deemed most likely, and tracks are constructed on the basis of lowest overall cost.

Fiji comes bundled with plug-ins for tracking, and we will use two of these for manual and automated tracking below.

▶ **Tutorial: Manual Particle Tracking**

The Fiji plug-in *Manual tracking* works well for tracking small numbers of objects over relatively few frames. This works really well for manually tracking nuclei of migrating cells, for example:

1. Open image stack in Fiji (see Note 1).

2. Click on *Plugins > Tracking > Manual Tracking*.

3. Verify that the image scaling is correct.

4. For purely manual tracking, keep "Centering Correction" unchecked.

5. To add the first track, click "Add Track."

6. Click on the image in the nucleus of interest, and repeat in the next frame.

7. If the movie ends, the track terminates; otherwise click "End Track."

8. Repeat adding new tracks until you are finished tracking all objects.

Note 1: If objects move slowly, the frame rate is too high, or you just want more coarse tracking, the number of frames to track can be reduced using *Image > Stack > Tools > Reduce*. Remember to verify the time scaling in the dialog box after reduction.

At this point the tracks can be viewed using the drawing function and/or the data from the "Results" window can be saved as a csv file. This file can be imported into R for further analysis. Manual tracking has the advantage that humans are pretty good at tracking objects, but if the number of objects to be tracked is large or the movies are long, an automated solution is the way to go.

▶ **Tutorial: Automated Particle Tracking**

TrackMate is an automated single-particle tracker for Fiji.[18] It has an intuitive graphical user interface (GUI) that guides the user through the tracking process. After opening the image stack in Fiji, click on *Plugins > Tracking > TrackMate*. Navigate through the dialogs using the "prev" and "next" buttons. As a default, TrackMate uses a Laplacian of Gaussian (LoG)[†] method to detect particles in each image (other methods are available). Specifying an estimated blob size and suitable offset is important to get good results. Scroll through the stack to check how TrackMate is detecting particles in early and late frames in the stack.

[†] Laplacian of Gaussian is a two-step method to find spots in images. A Gaussian filter is first applied before a Laplacian filter finds regions of contrast in the image.

The next important stage is to determine how TrackMate will link particles as a track. The user can specify gap-closing parameters that determine how far the search for linkages will extend in space and time. It is best to minimize these settings to prevent erroneous linkages. The exact settings need to be determined empirically, depending on the data set that you want to analyze. Finally, the tracks can be extracted for further analysis in R, although there is a sophisticated display of particle track data provided in TrackMate itself.

Kymographs

In cell biology, a kymograph is an image in which the pixel value along a line is plotted versus time. The name kymograph comes from the medical recording device that plots a reading (e.g., body temperature) over time using a rotating drum. Kymographs are very useful for visualizing processes like the movement of proteins along microtubules or vesicles along a neuronal process. One can readily see the number of events, their direction, speed, persistence, and intensity profile in a kymograph. They can be generated by selecting a line on an image stack and projecting the fluorescence over time (Fig. 4.8).

▶ **Tutorial: Generating a Kymograph**

1. Open the image in Fiji.
2. Project the movie into a new 2D image by using *Image > Stacks > Z-project...* and selecting *Sum Slices* (see Note 1).
3. Use the segmented line tool to draw a line along an interesting region.
4. Add this to the ROI Manager and save the ROI for reproducibility.
5. Transfer the ROI to the image using the Manager or by clicking "Reselect."
6. Generate the kymograph by selecting *Image > Stacks > Reslice* or by pressing the backslash key (\, see Note 2).

Note 1: With sum slices, the sum of all pixel values in the stack at their *x,y* location is displayed, giving a view of all places where the protein traveled during the movie. Alternative methods are "Average Intensity," in which the Sum Slices image is divided by the number of slices to give a mean value at each location, or "Maximum Intensity," in which the maximum value of all slices at the *x,y* location gives an intensity-compressed view of the stack. Other options are available.

Note 2: This command actually reslices the image stack along the selected line, so that if the stack is a z-stack rather than a time series, the resulting image is a cross section through the stack along this line. For kymographs, the convention is for distance to be shown horizontally with time shown vertically.

Frame 1   Sum Slices   Kymograph

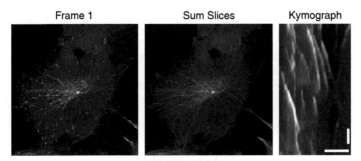

FIGURE 4.8. Kymograph of growing microtubule ends.
A confocal movie of an RPE1 cell stably expressing tdTomato-EB3, a protein that binds to the growing ends of microtubules. The first frame of the movie (Frame 1) or a projection of all frames (Sum Slices) is shown. A kymograph of EB3 comets traveling along one of these tracks is shown on the *right*. Scale bars, 1 μm horizontal; 5 sec, vertical.

An alternative trick to identify interesting regions of a stack is to use the image calculator function of ImageJ. With this GUI it is possible to take two images or stacks and perform simple operations with them (e.g., subtraction). If we have a movie of 10 frames and we want to look at frame-to-frame differences:

1. Duplicate frames 1–9 to create a new image (A).

2. Duplicate frames 2–10 to create a new image (B).

3. Using *Image Calculator*, subtract A from B. Select New Window and 32-bit Result.

The result will be a largely gray image (in which nothing has changed). Areas where the protein has moved *to* are white, and areas where the protein has moved *from* are black. Rather than a projection the result is a nine-frame movie (in this example) and can be animated to show the dynamic regions.

In the kymograph in Figure 4.8, the fluorescent signals appear as diagonal streaks. This is because the signal is moving along the line used to generate the kymograph. It is possible to measure one of these events with the help of a simple macro. Using the straightline tool, draw a line on the kymograph from the beginning to the end of the event. Now run this macro, which will calculate the velocity of the object from the selection that you have specified. Note that pxsize and tstep need to be set manually in the macro text, and that they depend on the scaling of the kymograph.

```
1   // set pxsize and tstep manually
2   // pxsize should match the original image
3   var pxsize = 0.069; // xy size in um
```

```
4    // tstep is the duration of the movie divided by the height of the
     kymograph in pixels
5    var tstep = 0.1; // seconds
6
7    // get the start and end points of straight line selection
8    getLine(x1, y1, x2, y2, lineWidth);
9    print("start ("+x1+" , "+y1+") - end ("+x2+" , "+y2+") ");
10
11   // calculation of speed.
12   dx = abs(x2-x1);
13   dy = abs(y2-y1);
14   dx *= pxsize;
15   dy *= tstep;
16   velocity = dx/dy;
17   print(dx+" um in "+dy+" sec");
18   print("Velocity (um/s)="+ d2s(velocity, 3));
```

Using this macro on the first object in the kymograph (Fig. 4.8), the output is

```
1    start (97 , 21) - end (136 , 188)
2    2.691 um in 16.7 sec
3    Velocity (um/s) = 0.161
```

Colocalization

Fluorescence microscopy can be used to infer whether two proteins are associated. Images of each protein tagged with a different fluorophore can be collected and then assessed. Demonstrating whether two proteins "colocalize" using unbiased image analysis is surprisingly challenging. Two distinct but overlapping approaches are used. First, *correlation* asks whether the signal intensities are related to one another. Second, *co-occupation* (or co-occurrence) determines if the signals corresponding to each protein are spatially related. Collectively this line of investigation is called "colocalization."

Colocalization is limited by resolution. At lower magnifications, determining whether protein A is expressed in the same cell type as protein B on a tissue scale is also referred to as colocalization. The same analysis methods can also be applied in this case, but the interpretation is very different.

Here, we are concerned with subcellular colocalization. Note that the resolution of light microscopy is too poor to determine whether protein A actually binds to protein B. They may just as easily both be bound to a third protein or simply be bound to the same organelle and not physically interact with one another at all.

The main difficulty with colocalization is that researchers want a single statistic that reports "colocalization." The reality is that neither the *co-occupation* nor the *correlation* methods work well for all cases and for all colocalization questions that might be asked. Most questions benefit from the generation of more than one statistic, and this leads to confusion over what is best practice.

The primary method for looking at *correlation* is to use Pearson's correlation coefficient (PCC). Dating from the nineteenth century, this method measures the strength of the linear relationship between pairs of intensity values. PCC is not affected by imaging parameters such as gain and offset, so it is fairly robust. Its main weakness is that if two signals are truly correlated but the intensities are not linear, PCC underestimates the association.

A related approach is to use Spearman's rank correlation coefficient (SRCC). Here, instead of using the raw intensities to calculate the strength of the relationship between two signals, the rank intensity is used. The advantage that SRCC can detect correlation where PCC falls short is reason enough to use SRCC over PCC by default.

A method that aims to look at both correlation *and* co-occupation is the overlap coefficient. This uses information from pixels that are not 0 and hence only looks at correlation where there is co-occupation. This method is not sensitive to camera gain but is sensitive to offset.

The primary method for examining co-occupation is to use Manders' coefficient.[19] Two statistics are produced (M_1 and M_2). These are the proportion of the total intensity of one channel that co-occupies the intensity of the other channel and vice versa. The split overlap coefficient (giving k_1 and k_2) is a related method.

Co-occupation coefficients tell us something different from correlation coefficients. Imagine that in the two channels of our image, we have one green spot completely within a larger red spot, such that there are no green pixels that are not red, but many red pixels that are not green. Using Manders' approach we can state absolutely that 100% of the green signal is red, but that some lower percentage of the red signal overlaps with the green signal.

There are a number of other analysis methods, such as the correlation method DeBias[20] and spatial methods such as nearest neighbor. The search for a one-size-fits-all approach to colocalization means that new methods will continue to be developed. Plug-ins for ImageJ will calculate a number of colocalization statistics. Some will even warn you if the method you have selected is not appropriate for the images you are working on. An excellent plug-in for ImageJ is JaCoP, coloc2 is packaged with Fiji, and there are many others.

Before attempting colocalization analysis, it is important to check the microscope setup used to capture images. If there is bleed-through from one channel to another, or if the signals from each fluorophore are not in register, then any colocalization method will give a flawed result. Imaging fluorescent beads or samples with single fluorophores can be used to test that the system is set up correctly.

If the experiment is to look at two proteins, then images of each protein together with an unrelated protein that does or does not colocalize are useful to get a feel for what the coefficients mean. Computational controls usually involve randomizing the images. This randomization can be done blockwise (Costes' approach) or at the level of individual pixels. One channel can simply be rotated by 90°, if the images are square. For time series, an effective strategy is to rearrange the order of images in one channel such that no channel pairs are coincident in time (a method known as derangement). This works well for images in which the proteins are located in puncta that are mobile, and less well for static structures. The purpose of these experimental and computational controls is to give confidence that the colocalization analysis is giving meaningful results.

The table below shows some guidelines for typical coefficient values indicating colocalization. However, the approaches work best as indicators of colocalization rather than absolute quantification methods. This means that calculation of PCC over time for two channels of a movie works well to look at relative changes in correlation.

Coefficient	Colocalization	Absence of colocalization
PCC	0.5 to 1.0	−1.0 to 0.5
SRCC	0.5 to 1.0	−1.0 to 0.5
Overlap coefficient	0.6 to 1.0	0 to 0.6
Manders' coefficient	>0.5	<0.5

▶ **Tutorial: Using R to Measure Colocalization over Time**

A live-cell imaging experiment has followed two proteins—one labeled with GFP, the other with mCherry (Fig. 4.9). A drug changes the localization of one protein so that it colocalizes with the other. What does this look like if we analyze the *colocalization* of those two signals?

We can take a 100 × 100-pixel ROI and plot the intensities of each channel at every *x,y* location.

If we randomize the pixels in channel 1 and replot, the result is very different.

Rather than using a plug-in in Fiji, we will analyze colocalization directly in RStudio, using this script. The output from R—using base graphics—is shown in Figure 4.10. This is an example of a *pipeline*, in which everything is done in R (see page 57).

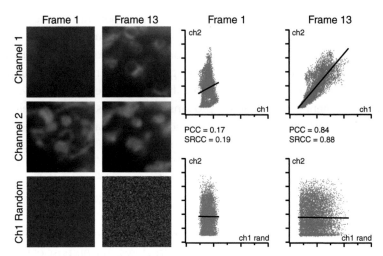

FIGURE 4.9. Studying colocalization of two proteins over time.
As described in the tutorial, the GFP-tagged protein (Channel 1) becomes colocalized with the mCherry-tagged protein (Channel 2) after addition of a drug (compare Frame 1 and Frame 13). Randomization of Channel 1 provides a control for colocalization.

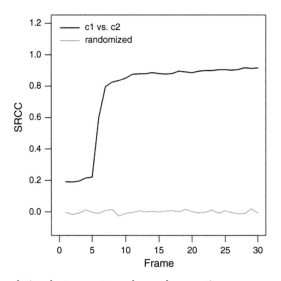

FIGURE 4.10. Correlation between two channels over time.
The result of the code example shows SRCC plotted for each frame of the movie. Channels 1 and 2 are compared for the data (black line) and for a case in which their pixels have been randomized (gray line).

```
1     # import each channel using EBImage
2     library(EBImage)
3     c1 <- readImage("c1.tif")
4     c2 <- readImage("c2.tif")
5     # assign the number of frames to a variable
6     fr = dim(c1)[3]
7     # make empty vector to take the SRCC results
8     r1 <- rep(NA, fr)
9     r2 <- rep(NA, fr)
10    # use a for loop to get for each frame, SRCC for c1 vs c2
11    for (i in 1:fr ) {
12      r1[i] <- cor(as.vector(c1[,,i]), as.vector(c2[,,i]), method=
          "spearman")
13      r2[i] <- cor(as.vector(c1[,,i]), sample(as.vector(c2[,,i])),
          method="spearman")
14    }
15    # plot out the result
16    png("corr.png")
17    par(cex = 1.5, mar = c(4, 4, 1, 1))
18    plot(0,0, type="l", xlim=c(0,fr), ylim=c(-0.1,1.2), xlab=
      "Frame", ylab="SRCC")
19    lines(1:fr, r1, lwd=2)
20    lines(1:fr, r2, col="gray", lwd=2)
21    legend("topleft",legend=c("c1 vs c2", "randomized"), bty = "n",
      col=c("black", "gray"), lwd=2)
22    dev.off()
```

GETTING THE RIGHT DATA OUT OF THE IMAGE

The golden rule of experimental data is that if you cannot see something by eye, it probably is not there. This is true of image analysis. Our eyes are actually very good at discerning images, and if there are patterns in the data that cannot be seen by eye, it is unlikely that a computer will be able to recognize them. The flip side of this is that if the computer analysis indicates that there is a change, you should also be able to see it by eye. If your analysis fails the eyeball test, you need a rethink. It could be the workflow that is at fault, or it could be that the images are simply not good enough for the analysis you have in mind.

Automation is great at removing human interference in the data analysis process, getting rid of bias and other distortion. However, it should not divorce the analysis from human judgment. When the results are spat out of the computer, what do they say? Do they make sense? Are all the controls giving the expected results? If the analysis says there is a fivefold change, does that look right to you? Is it what you expected? Is fivefold too big or too small? Can you see any evidence of artifacts, bleaching, or drift? Question it all.

It is important to validate the information you get from a workflow. In the case of new workflows, it is essential that they be validated fully.

Validation

How results of an analysis are validated depends on the workflow. Generally, the workflow should work as expected on some test images. These images are often referred to as "ground truth" images. They can be a prevalidated set of images or even a synthetic set of data in which key parameters are known. For example, imagine your workflow contains a step in which fluorescent puncta in a cell are detected using a thresholding algorithm. Some test images can be generated with a known number of puncta (plus noise) and then used to test how good the detection algorithm is. If detection fails on the test images, there is no way it will work on the real data. It is important to test all major steps in an analysis workflow in this way to ensure they are fully functional (see Write Modular Code in Chapter 6).

Where do you stop, and how much validation is necessary? Established software like R has been extensively road-tested. Chances are that any errors have been introduced by the code that you have written. You are unlikely to find a bug in the core functions of R. You can also assume that the default options of most built-in functions are optimized for widespread use. ImageJ/Fiji, on the other hand, is still under active development, which means that it is susceptible to bugs. They get ironed out quickly when reported, but you should bear in mind that an error you encounter could be due to the way this version of the software works. If you are incorporating someone else's code or plug-in into your workflow, you need to validate that output and ensure you understand what it does in order to fix it if something goes wrong.

Back to Square One

Sometimes, things do not go as planned. You have done an experiment; you are sure that there is an effect, but your analysis cannot determine one. The workflow is working properly and there is no other explanation. This can mean that you just need to go back to square one and acquire some new images. Look again at the experimental setup and determine if the dynamic range of the instrument is sufficient or if there is any other technical reason (perhaps bleed-through or drift) the effect is obscured.

If the images are clear and further improvements are not possible, then your analysis approach may be suboptimal in some way. Go back, not to square one, but somewhere after that, and figure out if improvements can be made. It is important to realize that the problem may be intractable in the sense that further investment of time and effort cannot justify the result at the end.

◆ *Golden Rules*

- Location matters. Where you place the ROI or how you segment the image will affect your results.
- If you cannot see something by eye, it probably will not be revealed by a computer.
- Accurate segmentation is key to most image analysis pipelines.
- Automating segmentation can be difficult; effort and experimentation are needed to crack challenging segmentation problems.
- Any image analysis step needs to be tested rigorously: Use synthetic data to test your workflow.

CHAPTER 5

Statistics

Mathematical literacy tends to be lower among cell biologists than in some other fields. Theories abound as to why, and one that I favor is this: Cell biologists simply got out of the habit of thinking quantitatively during the molecular biology era. Experiments in which a gene is deleted and all the cells die do not really need quantification. This style of science is summed up in the famous quote by Rutherford: "If your experiment needs statistics, you ought to have done a better experiment."

It has led to a regrettable loss of interest in quantification and statistics, but this must change.

Experiments are difficult and lab work is time-consuming. Tackling the dry matter of quantification tends to take a backseat when there is challenging lab work to perform and important data to collect. However, it is no good being able to perform very clever experiments and collect high-quality data if you cannot analyze that data yourself.

There's a further problem: Statistics textbooks can be confusing, especially for a cell biologist. These books are aimed at mathematicians and statisticians and focused on the detail of calculations, or they are pitched at biomedical scientists but emphasize experiments using patients, which are unlike those encountered in the cell biology lab. The aim here therefore is to give an overview of key concepts in experimental design and statistics that are relevant to cell biology.

DESIGNING AN EXPERIMENT

There are four things to consider when designing an experiment.

Replication: This is a core principle in science. If $n = 1$, the conclusions that can be drawn are severely limited. The question of "What is n?" is discussed below.

Appropriate controls: Ideally an experiment tests the effect of changing one variable while controlling for all others. This allows us to conclude that any changes are due to altering that variable alone. Negative and positive controls should be included in every experiment. A negative control is one in which no effect is expected, and a positive control is one in which the effect is known (or predicted) to occur.

Statistical power: As we will see below, the ability to interpret your experiment depends on power, which in turn is linked to the sample size needed for your experiment as well as the effect size.

Reduce error and eliminate bias: Identify potential sources of error and bias in your current experimental design. There may be errors in the measurement you take, in the sampling procedure, or in the experimental manipulation. Look hard at the accuracy of the measurements you make. How replicable are the measurements? Are you sampling randomly? Is it possible that cells in one group have been treated differently from those in another? Elimination of bias can be difficult. The analyst should be blind to the experimental conditions in any case in which there are human decisions to be made (e.g., manually drawing around cells). A simple way to reduce bias is to automate the analysis as much as possible, because computers are not susceptible to wishful thinking.

WHAT IS *n*?

This is a deceptively simple question. The answer is often complex, and straightforward advice is therefore difficult. There is a conflict between what biologists may think is important and statistical reality. This question is further muddied by inconsistent terminology. Sometimes people talk about replicates and repeats, the meanings of which can differ. Others talk about technical and biological replicates. Which one of these (if any) refers to pseudoreplication is not clear. Here are some definitions to get us onto the same page[21,22]:

Biological unit	This is the entity about which we will make conclusions.
Experimental unit	This is the entity that is randomly assigned to the different experimental conditions; it determines *n*.
Observational unit	This is the sample of measurements that are taken.

Note that *n* refers to the sample size, whereas the population size is *N*. The population is usually a theoretical entity and an enormous number. Most of the time we talk about *n*, as we are considering the sample population.

Here is an example to think about. We want to test if protein x is a modulator of macropinocytosis. To do this, we use EGF stimulation of HEK293 cells in which we have depleted a protein by using RNA interference (RNAi), including appropriate controls. We incubate the cells with a fluorescent dye that is taken up by macropinocytosis and use microscopy to measure the number of fluorescent puncta inside each cell. Let's put some numbers on this and ask, "What is *n*?" Starting at the analysis, we count the number of puncta in 20 cells per condition. The cells are plated out and transfected once. So what is n? The answer is 1. It is not 20. It is 1 for each condition. Suppose we prepare four wells of cells and image five cells from each well per condition to image 20 cells in total per condition? What is n? The answer is still 1 per condition. Any answers greater than 1 refer to pseudoreplication. The only way to increase n is to repeat the entire experiment.

Using the terminology above:

Biological unit	EGF-stimulated HEK293 cells	1
Experimental unit	Number of times experiment is performed	1 (in each condition)
Observational unit	Number of measurements taken	20

What you want to know is whether the experimental manipulation had an effect that was statistically significant. That is, you changed one variable and what effect, if any, did it have? So the correct test is at the level of the experimental unit. In cell biological experiments, the experimental unit is often not a single cell. Cell lines are highly homogeneous and what you really want to know is whether the manipulation is reproducible (i.e., do you see the same effect if you repeat the experiment)?

It is possible to make comparisons at lower levels, but they tell you something different. In the example above, we can test if the depletion of the protein changed macropinocytosis with $n = 20$, but this would not tell us how generalizable the result is. Any significant effect detected would depend on the EGF stimulation in that well of the plate, on how well the protein was depleted that day, and so on. It is only by repeating the experiment that you begin to get a feel for what would happen in someone else's lab were they to repeat your experiment. And this is what you want to know.

Comparisons at yet lower levels are increasingly meaningless. For example, we could measure the brightness of each of the vesicles and compare cell 1 and cell 2, but who cares? Comparisons of the observational unit and below tell us about variability between cells or about your measuring error on that day, but these comparisons are not worth reporting.

Single cells are not usually considered the experimental unit. In the example above, EGF stimulation and fluorescent dyes were applied to the whole well. Because we are using a cell line, it is unlikely that responses will differ between two cells in the same well. In addition, any experimental error—for example, applying too much EGF—would affect all cells in the well. You might think that it is justified to split out the cells into different wells and to perform the experiment in parallel. Again, the results here are not as generalizable as repeating the experiment on different days. The important thing to remember is that each n must be independent. In this example, we need to repeat the experiment to increase *n*, and for each repeat we can take the mean of the observational units for each treatment and use those for comparison.

Sometimes a single cell can be the experimental unit. This is the case if treatments are applied directly to each cell *independently* and a recording is taken from that cell *in series*. An example here is a fluorescence recovery after photobleaching (FRAP) experiment. A single cell is imaged, a ROI is bleached, and the recovery of fluorescence is recorded. Next, another cell from a new field of view or new dish is imaged in the same way, and *n* increases by 1. It is still advisable to repeat the experiment several times (e.g., *n* = 15, with five cells from three different experiments). It is worth assessing if any differences between experiments can be detected in the results, but otherwise the cells can be treated as independent units. If there are differences between experiments—for example, if the control values are consistently lower (across all conditions) for one experiment—then normalization may be required. Fold changes relative to the control can be used rather than absolute values. If the recordings are taken in parallel, the cells can still be considered independent, but again testing for batch effects is prudent. Nonetheless, the experimental unit and observational unit are both 15.

Unfortunately, it is not possible to draw conclusions beyond the experimental unit. In the example above, we can draw conclusions about one cell line and one particular form of macropinocytosis (the biological unit). To make more general conclusions about human cells or about all forms of macropinocytosis, we would need to repeat the experiments on other cell lines and/or using different methods. Currently, most cell biologists prefer to replicate their work on the same cell line (biological unit) and acknowledge the limitations of their conclusion.

Two terms that have emerged in proteomics and other areas are "technical" and "biological" replicates. These terms are not used consistently, but they often refer to a complete rerun of an experiment (biological replicate) or to a pseudoreplication (technical replicate). An example of pseudoreplication is where the sample has been divided in two and then analyzed twice. There are sometimes valid reasons to do this, perhaps to reduce noise or variability in the measurements, but it is important to recognize that they are not true replications. Pseudoreplication can creep into experimental design because it is significantly easier to "replicate" this way than to actually replicate the experiment.

In the end, determining *n* is a judgment call. Remember that the various statistical tests were developed to prevent you from making a fool of yourself by claiming an effect when what you have observed was due to chance.[23] Falsely inflating *n* increases the chance that you will make a mistake and that others will not be able to repeat your work. Opinions will differ on whether pseudoreplication has occurred, but tending toward more rigorous experiments will always make your results more believable.

There are some excellent guides to experimental design that are worth consulting if you are not sure how to define the biological, experimental, and observational units in your experiment.[21,22] These resources also discuss randomization, blinding, blocking, and other techniques that are not covered here.

◆ *Golden Rules*

- Good experimental design is essential: Use controls and random sampling and eliminate bias.
- Repeat your experiments rather than collecting more data from a single experiment.
- Be aware of pseudoreplication and be able to recognize and correct for it.

Why Does *n* Matter?

n matters because it gives us confidence that the sample accurately reflects the population. The more independent measures you take from a population, the more accurate the picture of the population is, based on your sample.[*]

The minimum size of *n* is relative to the effect observed in your experiment. A huge effect means a small *n* is sufficient. Conversely, smaller effects require larger numbers of samples to be accurately detected. In cell biology, *n* is usually small, and it is safe to promote a "more *n* is good" message. If pseudoreplication has occurred or a technique is being used that generates huge *n* data, there are pitfalls that you should be aware of. For example, trivial differences can be attributed to very small *p*-values when *n* is very large. *p*-values are defined and discussed in more detail at the end of this chapter.

Power Analysis for Cell Biologists

Power analysis is a statistical method to determine the sample size needed. At this point, most statistical guides begin to discuss clinical trials or animal experiments. This is because *n* in this context has financial and ethical implications: Making more observations has very real consequences. In cell biology we are not so concerned about this. Cells in culture are easy to grow cheaply, and doing more

---

[*] The more technical view is that *n* matters because in parameter estimation, the number of pieces of information that go in are defined as "degrees of freedom," and the number of degrees of freedom is $n - 1$.

replications is usually no big deal. This low barrier to cell biology experiments is one reason why power analysis is not a core part of our experimental design. In a clinical trial, power analyses are crucial to getting any meaningful results from a trial. They need to be performed up-front and approved before any work can commence. Many scenarios are modeled to ensure that the trial does not fail because of poor experimental design. In cell biology, there is no real need to do this. A pilot experiment can be done in a few days. Likely outcomes can be quickly determined, and experimental design or sample sizes can be adapted— for example, by repeating the experiment with altered conditions if anything needs optimization.

In null hypothesis statistical testing, the goal is to accept or reject the null hypothesis (H_0), which is usually a statement that the experimental manipulation has no effect. There are actually four possible outcomes here. The hypothesis is accepted or rejected, correctly or incorrectly. There are special terms given to the two error outcomes:

	H_0 is true	H_0 is false
H_0 is rejected	Type I error	No error
H_0 is not rejected	No error	Type II error

A useful way of remembering which error is Type I or II is to remember the story of "The Boy Who Cried Wolf." At the beginning he claimed there was a wolf when there was not (Type I error, false-positive), and in the end no one believed there was a wolf when there was one (Type II error, false-negative). The probabilities of these two error types have special terms, α and β, respectively:

	H_0 is true	H_0 is false
H_0 is rejected	α	$1 - \beta$
H_0 is not rejected	$1 - \alpha$	β

If H_0 is *not* rejected, there are two possible reasons: Either the null hypothesis is correct or the sample size is not sufficient to do a powerful enough test.

Power refers to the sensitivity of the statistical test. It is the value of $1 - \beta$ and is commonly set at 90% (0.9), which gives good sensitivity without demanding huge sample sizes. The significance level, α, is usually set at 0.05. This value relates to p-values used in statistical tests (discussed later in this chapter). Now, if we also know the effect size (ES) and the variability (σ^2), we can calculate the sample size requirement. These parameters are all interrelated,

$$\text{Power} \propto \frac{ES\alpha\sqrt{n}}{\sigma^2}, \tag{5.1}$$

so you can calculate the minimum effect size that can be detected if you know the sample size, and so on. To give a simple example, suppose that we have previously measured the cytosolic fluorescence of an mCherry-tagged nuclear protein released in response to a compound. The control compound results in mean fluorescence of 4000 units with a standard deviation (σ) of 1600 in 20 different experiments. If we wanted to be able to detect a 50% increase in fluorescence, how many experiments do we need to do? So the effect size we would like to detect is a change (Δ) of 2000 (i.e., 50% increase giving a mean of 6000). Assuming the standard deviation increases proportionally (to 2000), how many samples (in each group) would we need at $\alpha = 0.05$, $\beta = 0.9$? Try to calculate the answer using power.t.test() in R:

```
1   power.t.test(n = NULL, delta = 2000, sd = 2000, sig.level = 0.05,
    power = 0.9)
```

This gives $n = 14.5$. Because it is not possible to do half an experiment, it is common to round upward. So we need to perform 30 experiments, 15 with the control drug and 15 with the test compound.

Other calculations are possible. For example, if the compound is very precious and only nine experiments (in each group) can be done, you can calculate the magnitude of change that can be detected at $\alpha = 0.05$, $\beta = 0.9$ and a similar standard deviation:

```
1   power.t.test(n = 9, delta = NULL, sd = 2000, sig.level = 0.05,
    power = 0.9)
```

Now $\Delta = 3257.6$, so large changes of >81% can be detected with this sample size.

In these examples, we had the information required to do the calculations. If a parameter is unknown, the calculations can be done using parameters from a similar data set or for a range of values in order to give an estimated answer.

BASIC STATISTICS THAT YOU WILL NEED

In this section, we'll use a menu approach to picking an appropriate statistical test for your experimental data (see Table 5.1). This advice is intended to give you clear guidance on which methods are *inappropriate* and should be avoided. However,

TABLE 5.1. *Recommended statistical tests for different data types*

What do you want to do?	Measurement (normal distribution)	Measurement (non-normal) or rank	Binomial (two outcomes)
Describe one group	Mean, SD (σ)	Median, interquartile range (IQR)	Proportion
Compare one group to a value	One-sample *t*-test	Wilcoxon test	Chi-squared
Compare two unpaired groups	Unpaired *t*-test Permutation test ($n > 10$)	Wilcoxon–Mann–Whitney two-sample rank test	Fisher's exact test or chi-squared
Compare two paired groups	Paired *t*-test	Wilcoxon signed rank test	McNemar's test
Compare three or more unmatched groups	One-way ANOVA	Kruskal–Wallis test	Chi-squared test
Compare three or more matched groups	Repeated-measures ANOVA	Friedman test	Cochran's Q test
Quantify association between two variables	Pearson correlation	Spearman correlation	
Predict value from another measured variable	Simple linear regression	Nonparametric regression	Simple logistic regression
Predict value from several measured or binomial variables	Multiple linear (or nonlinear) regression		Multiple logistic regression

Modified from Table 37.1 in Motulsky 1995[25]; reproduced with permission of the Licensor through PLSclear.

you should recognize that statistical methodology in cell biology (as presented here) lags a few decades behind current statistical thinking. Knowledge gained here will put you at the forefront of numeracy in cell biology but would not impress an actual statistician. To delve deeper into the subject, there are some accessible detailed guides available.[24] With that caution out of the way, let's look at basic statistics.

To figure out what statistical test you need to do, look at Table 5.1. But before doing that, you need to ask yourself a few things:

- What are you comparing?
- What is *n*?
- What is your hypothesis?
- What will the summary statistic tell you?

If you are not sure about any of these things, whichever test you do is unlikely to tell you very much. To pick the correct test from the table, you need to determine what type of data you have:

- Measurements: Most data you analyze in cell biology will be in this category. Examples are number of spots per cell, mean GFP intensity per cell, diameter of nucleus, speed of cell migration, etc.

  —Normally distributed: This means it follows a "bell-shaped curve," otherwise called a "Gaussian function."

  —Not normally distributed: Data that do not fit a normal distribution—skewed data, or data better described by other types of curve.

- Binomial data. These are data with two possible outcomes. A good example here is mitotic index measurement (the proportion of cells in culture that are in mitosis). A cell is either in mitosis or it is not.

- Other: Maybe you have ranked or scored data. This is not very common in cell biology. A typical example here would be the Likert scale used in questionnaires (1 = strongly disagree, …, 5 = strongly agree). For a cell biology experiment, you might have a scoring system for a phenotype—for example, fragmented Golgi (1 = is not fragmented, 5 = is totally dispersed). These arbitrary classifications are not a good idea, especially if the person scoring is not blinded to the experimental procedure. It is best practice to come up with an unbiased measurement procedure.

Most of the data in cell biology are *unpaired* or *unmatched*. Individual cells are measured and you have, say, 20 cells in the control group and another 20 cells in the test group (40 different cells in total). These are unpaired (or unmatched in the case of more than one test group) because the cells are different in each group. If you had the same cell in two (or more) groups, the data would be paired (or matched). An

TABLE 5.2. Refresher on summary statistics

Statistic	a.k.a.	Type	Details
Mean	Average, μ, $\bar{x}$ (sample), $E[x]$ (population)	Center	Measure of central tendency, $\bar{x} = \frac{1}{n}\left(\sum_{i=1}^n x_i\right) = \frac{x_1 + x_2 + \cdots + x_n}{n}$
Median	M, $\mu_{1/2}$, $\tilde{x}$	Center	Midpoint of a numerically ordered sample
Standard deviation	SD, s.d., s (sample), σ (population)	Dispersion Descriptive	Typical difference between the sample members and the mean, $s = \sqrt{\dfrac{\sum_{i=1}^n (x_i - \bar{x})^2}{n-1}}$
Standard error of the mean	SEM, s.e.m., $s_{\bar{x}}$	Dispersion Inferential	Measure of how variable the mean will be when the study is repeated, $s_{\bar{x}} = \frac{s}{\sqrt{n}}$
Confidence interval	CI, typically 95% CI	Dispersion Inferential	The range that you can be 95% sure contains the mean, $\left(\bar{x} - t*\frac{s}{\sqrt{n}},\ \bar{x} + t*\frac{s}{\sqrt{n}}\right)$, where t is the critical value $t_\alpha(df)$, where $\alpha = (1 - C)/2$ and $df = n - 1$; C is the confidence level
Range		Dispersion Descriptive	The spread of the data, x_{min} to x_{max}
Interquartile range	IQR	Dispersion Descriptive	How the central half of the data is spread around the median, Q_3 to Q_1, 75th to the 25th percentile; median of the upper half and lower half of sample, respectively
Median absolute deviation	MAD	Dispersion Descriptive	Robust, unweighted measure of dispersion; MAD = median ($|x_i - \text{median}(x)|$), for symmetric distributions MAD $= \frac{1}{2}$IQR

Other measures of central tendency exist. The mode is the most common value in a sample and there are other types of mean—for example, geometric mean.

There are other methods to calculate CIs—for example, using bootstrapping.

example of a paired data set would be one in which you have 10 cells that you treat with a drug. You take a measurement from each of them before treatment and a measurement afterward. So you have paired measurements: one for cell A before treatment, one after; one for cell B before, one after; and so on.

A Refresher of Summary Statistics

Before we take a look at how to perform statistical tests, let's remind ourselves of the common descriptive statistics used to summarize a sample. The terms in Table 5.2 describe the center and the dispersion of the data; these are known as the lower statistical moments. Higher moments, which describe the shape of the data, such as skewness and kurtosis, are not commonly reported.

Always Plot Out Your Data

Dealing with numbers can quickly become abstract. Plotting a data set allows us to see the shape of the data, and it can reveal much more than a series of statistical tests. In other words, simply looking at the data can reveal things that are not obvious from summary statistics. In a classic example, F.J. Anscombe showed that data sets with very similar summary statistics can actually be very different from one another. The simplest way to see these differences is to plot them out. Try this exercise in R, executing the following script line by line:

```
1    library(ggplot2)
2    library(gridExtra)
3    # load in the data
4    data(anscombe)
5    # look at summary statistics, note the similarities
6    summary(anscombe)
7    # look at the correlation between x and y pairs
8    sapply(1:4, function(x) cor(anscombe [, x], anscombe [, x+4]))
9    # variance in the data is very similar too
10   sapply(5:8, function(x) var(anscombe [, x]))
11   # do linear regression on the first pair x1 and y1
12   lm(y1 ~ x1, data = anscombe)
13   # shows that intercept is 3 and the slope is 0.5
14   # check the other data pairs using
15   # lm(y2 ~ x2, data = anscombe)
16   # now plot out the data out and look at differences
17   p1 <- ggplot(anscombe) + geom_point(aes(x1, y1))
18   p2 <- ggplot(anscombe) + geom_point(aes(x2, y2))
```

```
19   p3 <- ggplot(anscombe) + geom_point(aes(x3, y3))
20   p4 <- ggplot(anscombe) + geom_point(aes(x4, y4))
21   # the figure in the book is generated with more complicated code
22   grid.arrange(p1, p2, p3, p4)
```

Anscombe's quartet demonstrates that we can learn a lot about a data set that is otherwise hidden by summary statistics (Fig. 5.1). A good first step before doing any statistical tests is to plot your data and look at it.

Let's look at some more data. We'll make a random but reproducible data set and use base R graphics[†] to look at it (Fig. 5.2). Using R, follow the code examples in this section:

```
1    # 100 random values taken from the normal distribution
     (mean = 0, standard deviation = 1)
2    set.seed(1)
3    data0 <- rnorm(100)
4    # shows us the values in the vector, plotted as points
5    plot(data0,t="p")
```

In this example we want to know whether the data are normally distributed. Figure 5.2 shows that the points are clustered around 0, but it is difficult to see from this plot how the data are distributed. To simply look at this we can make a histogram of the values and see if the shape of the histogram is Gaussian (Fig. 5.3):

```
6    hist(data0)
```

This technique is good, although information in the tails of the distribution might not be visible. Another technique is a quantile–quantile (QQ) plot, in which the data percentiles are plotted against those for the equivalent normal distribution:

```
7    qqnorm(data0, xlim = c(-3,3), ylim = c(-3,3), main = "Normal Q-Q
     Plot",
8       xlab = "Theoretical Quantiles", ylab = "Sample Quantiles", plot.
        it = TRUE)
9    abline(a=0,b=1,lty=2)
```

[†] Very basic graphics can be generated in R using simple, easy-to-understand code. These core functions are referred to as "base R" to distinguish them from more advanced graphics generated with libraries such as ggplot2.

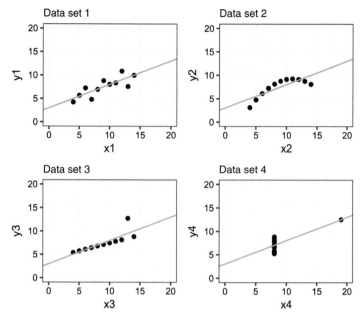

FIGURE 5.1. Anscombe's quartet.
Each data set is very different. Notice how the intercept and slope of the line-of-best-fit through each data set are the same. What did the summary statistics show?

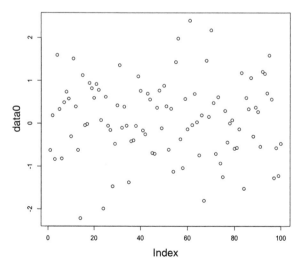

FIGURE 5.2. Point view of our data.

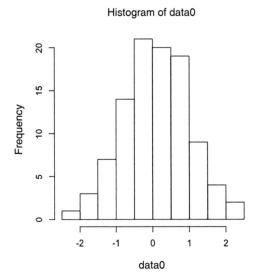

FIGURE 5.3. Histogram of our data.

If the data diverge from a line showing $y = x$, this indicates that the distribution is skewed (Fig. 5.4).

Finally, if the data set is small, it is possible to test if your data set is normally distributed. There are several tests (e.g., Kolmogorov–Smirnov, Jarque–Bera), but the easiest and most intuitive is Shapiro–Wilk:

```
10    shapiro.test(data0)
11
12            Shapiro-Wilk normality test
13
14    data: data0
15    W = 0.9956, p-value = 0.9876
```

We will look at *p*-values in detail later; however, this output is telling us that $p > 0.05$, which means that the distribution is indistinguishable from a normal distribution. Statistical tests based on common distributions (most notably the normal distribution) are called *parametric*, whereas procedures that work without those assumptions are called *nonparametric*. In cell biology, the numbers of measurements can be quite small. This means that although the population may be normally distributed, your sample taken from the population might not be. In these cases it is advisable to use nonparametric tests. We will cover these tests below.

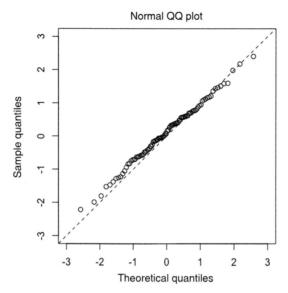

FIGURE 5.4. QQ plot of our data.

Descriptive Statistics

The purpose of descriptive statistics is to simplify or to summarize a data set. As we saw above, we can get the mean, median, quantiles, and minimum and maximum values by using the summary() function in R:

```
16   summary(data0)
17      Min.   1st Qu. Median   Mean   3rd Qu.   Max.
18    -2.2150  -0.4942  0.1139  0.1089  0.6915  2.4020
```

Other summary statistics, such as the standard deviation, can be accessed via specific functions—for example, sd() gives us the standard deviation:

```
19   sd(data0)
20   [1] 0.8981994
```

Summary statistics allow us to look at the properties of the sample that has been measured, rather than making any inferences about the population or comparing it with other samples.

STATISTICAL TESTS

In this section we will perform a series of statistical tests in R. We will use simulated data to represent experimental measurements made at the level of the *experimental unit*. Starting at the top of Table 5.1, we will progress to more complicated statistical tests. You can play along using the available R Scripts.

Compare One Group to a Value

It is unlikely that you will need to compare a data set with a reference value. In cell biology, we tend not to use hypothetical values for comparison; we almost always have experimental values from appropriate controls. However, to compare one set of measurements to a hypothetical population that has a mean of 1, for example, we use the t.test () function in R:

```
21    t.test(data0,mu=1)
22
23            One Sample t-test
24
25    data:   data0
26    t = -9.9211, df = 99, p-value < 2.2e-16
27    alternative hypothesis: true mean is not equal to 1
28    95 percent confidence interval:
29      -0.06933487  0.28710961
30    sample estimates:
31    mean of x
32    0.1088874
```

Compare Two Groups

You can use the Student's *t*-test with Welch's correction for normally distributed data when you have a test group and a control group and no other groups. You might use this to compare the speed of microtubule growth in drug-treated cells with untreated control cells. This is an unpaired test and we can do it using t.test () in R:

```
1    # some simulated data for example, n = 8 for each group
2    set.seed(1)
3    data1 <- rnorm(8, mean = 1, sd = 0.3)
4    data2 <- rnorm(8, mean = 3, sd = 0.9)
```

```
5    # data1 is control, data2 is drug-treated
6    t.test(data1,data2)
7
8       Welch Two Sample t-test
9
10   data: data1 and data2
11   t = -5.2874, df = 7.8403, p-value = 0.0007899
12   alternative hypothesis: true difference in means is not equal to 0
13   95 percent confidence interval:
14     -2.885973  -1.128771
15   sample estimates:
16   mean of x mean of y
17     1.039436  3.046808
```

In this example we had a sample of eight measurements of the experimental unit in each group. If $n > 10$, and the variances of both samples are equal, you can also consider a permutation test.

If your data are paired, the t.test() can be used with an additional parameter. We will reuse data1 and data2 to do this and assume that they in fact came from a paired procedure—for example, the microtubule growth is measured for each independent cell twice (before and after addition of a drug):

```
18   # For paired data, use the additional parameter.
19   t.test(data1,data2,paired=TRUE)
20
21      Paired t-test
22
23   data: data1 and data2
24   t = -5.4765, df = 7, p-value = 0.0009294
25   alternative hypothesis: true difference in means is not equal to 0
26   95 percent confidence interval:
27     -2.874108  -1.140636
28   sample estimates:
29   mean of the differences
30         -2.007372
```

The default in R is to do two-tailed *t*-tests. This tests the null hypothesis that the data are not different in either direction (i.e., not greater than nor smaller than the control). It is rare that normally distributed data should be tested with a one-tailed test. It is valid to do so if the alternative hypothesis is directional and the data from the test group can only move in one direction. Unless you are absolutely certain about this, it is advised to stick to two-tailed testing in all cases.

The nonparametric equivalent to the *t*-test is the Wilcoxon rank sum test, also known as the Mann–Whitney test. You might use this for data that are skewed, such as circularity measurements of vesicles in electron micrographs taken from control and RNAi-treated cells.[#] We can run this in R using the `wilcox.test()` function:

```
1    # some simulated data for example
2    set.seed(1)
3    data3 <-rbeta(40,5,1)
4    data4 <-rbeta(40,2,1)
5    wilcox.test(data3,data4)
6
7          Wilcoxon rank sum test
8
9    data:   data3 and data4
10   W = 1130, p-value = 0.0013
11   alternative hypothesis: true location shift is not equal to 0
```

If these data were in fact paired, we could use `wilcox.test()` again, with an additional parameter. Note how the syntax is similar to switching between unpaired and paired *t*-tests:

```
12   # For paired data, use the additional parameter.
13   wilcox.test(data3,data4,paired=TRUE)
14
15          Wilcoxon signed rank test
16
17   data:   data3 and data4
18   V = 624, p-value = 0.003358
19   alternative hypothesis: true location shift is not equal to 0
```

[#] Distributions of circularity measurements ($4\pi[\pi r^2/(2\pi r)^2]$) of approximately circular objects are skewed because the measurement must be (0, 1]. There will be a low density approaching 0 and a very high density up to and including 1, with no measures beyond this cutoff.

Another useful nonparametric method to compare two unpaired groups is the Kolmogorov–Smirnov test. This test asks whether the test group can be described by a similar distribution to the control. Presented together with a cumulative density plot of the data, the result of this test is very intuitive—similar to the QQ plot described earlier. This type of test and presentation is common for data sets in which the time for a cell to reach a milestone has been measured—for example, time to detach from the substrate. We will reuse our two data objects from the previous test, and the `ks.test()` function in R:

```
20    # Kolmogorov-Smirnov test comparing data3 and data4
21    ks.test(data3,data4)
22            Two-sample Kolmogorov-Smirnov test
23
24    data:   data3 and data4
25    D = 0.4, p-value = 0.003018
26    alternative hypothesis: two-sided
27
28    # generate the cumulative distribution functions
29    cdf3 <- ecdf(data3)
30    cdf4 <- ecdf(data4)
31    # open a pdf file
32    pdf("ksplot.pdf")
33    # plot out the result
34    plot(0, 0, type="l", xlim=c(0,1), ylim=c(0,1),
      xlab="Circularity", ylab="Cumulative probability")
35    lines(cdf3, lwd=2)
36    lines(cdf4, col="gray", lwd=2)
37    legend("topleft",legend=c("data3", "data4"), bty = "o", col=c
      ("black", "gray"), lwd=2, bg = "white")
38    dev.off()
```

First, plotting out the data shows us how skewed the data sets are (Fig. 5.5). If the data were normally distributed, the cumulative density functions would look S-shaped. Second, the two distributions are clearly different from one another. The Kolmogorov–Smirnov test gives us a *p*-value of 0.003, indicating that the distributions are distinct. Note that this approach can also be used with data that follow a normal distribution.

Instead of measurements on a continuous scale, your data might belong to discrete categories. A classic example is data with binary outcomes—for example,

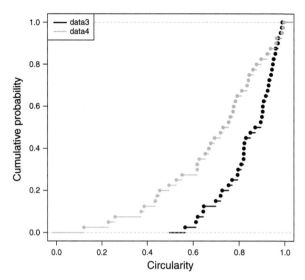

FIGURE 5.5. Cumulative density function for two experimental groups.

what proportion of cells are in mitosis or interphase or the mitotic index of a cell population. You can use a 2 × 2 contingency table and chi-squared text for independence. In the example below we will construct a 2 × 2 contingency table. Have a look at it and then use `chisq.test()` to do the test:

```
1    data5 <- matrix(c(18, 460, 57, 466), byrow = TRUE, 2, 2)
2    rownames(data5) = c("Ctrl","RNAi")
3    colnames(data5) = c("Mitosis","Interphase")
4    data5
5        Mitosis Interphase
6    Ctrl     18      460
7    RNAi     57      466
8    chisq.test(data5)
9
10          Pearson's Chi-squared test with Yates' continuity
     correction
11
12   data: data5
13   X-squared = 17.318, df = 1, p-value = 3.161e-05
```

All of the tests above apply to two experimental groups. If you have more than two groups, you should not do multiple versions of these tests. Correction for multiple comparisons is required; otherwise the probability of a Type II error increases.

Comparing Three or More Groups

For normally distributed data, you need to do a one-way ANOVA followed by a post hoc test to compare more than two groups. The ANOVA will tell you if there are any differences among the groups, but it will not tell you which ones are different. You can discern which pairs are significantly different by doing a post hoc test. Several tests exist—for example, Dunnett's test, which is useful if you have one control value and a bunch of test conditions. Tukey's post hoc comparison, which I favor, looks at all combinations of groups and as a result it is more conservative than Dunnett's test.

Imagine an experiment in which you have measured GFP fluorescence in a control RNAi group versus two other conditions. In R, let's build a data frame called data6 that has two columns, "gfp" and "rnai." These columns will contain our measurments and the experimental group the measurement comes from, respectively:

```
1   data6 <- data.frame(
2       gfp = c(294.9, 287.6, 195.2,
3           290.3, 309.7, 201.2,
4           291.1, 322.7, 203.7,
5           325.7, 301.7, 205.5,
6           302.2, 310.1, 196.4,
7           303.8, 297.5, 211.8,
8           297.1, 298.2, 209.8,
9           287.9, 292.5, 201.3,
10          311.9, 312.2, 191.5,
11          294.8, 284.8, 195.9),
12      rnai = factor(rep(c("ctrl", "sirna1", "sirna2"), 10)))
13  # have a look at the data frame
14  View(data6)
```

Next, we will perform the ANOVA and check the result. We can do this using the aov() function. The arguments for the function specify the data frame (data6) and the columns we want to compare. Here, gfp ~ rnai means do changes in the "gfp" measurements depend on the "rnai" variable?:

```
15    # do ANOVA
16    data6_av <- aov(gfp ~ rnai, data6)
17    summary(data6_av)
18                 Df Sum Sq Mean Sq F value Pr(>F)
19    rnai          2  66156   33078   310.2  <2e-16  ***
20    Residuals    27   2880     107
21    - - -
22    Signif. codes:  0 '***' 0.001 '**' 0.01 '*' 0.05 '.' 0.1 ' ' 1
```

The output tells us that the significance is "***." This means the *p*-value is between 0 and 0.001, and if we look at the table we see it is $<2 \times 10^{-16}$. What this tells us is that a difference that depends on the "rnai" variable has been found. We can now use Tukey's HSD (honestly significant difference) post hoc test to see what differences there are between the three RNAi conditions:

```
23    # do Tukey test
24    data6_tukey <- TukeyHSD(data6_av)
25    # look at results
26    data6_tukey
27        Tukey multiple comparisons of means
28          95% family-wise confidence level
29
30    Fit: aov(formula = gfp ~ rnai, data = data6)
31
32    $rnai
33                        diff        lwr        upr       p adj
34    sirna1-ctrl         1.73   -9.721109   13.18111   0.925785
35    sirna2-ctrl       -98.74 -110.191109  -87.28889   0.000000
36    sirna2-sirna1    -100.47 -111.921109  -89.01889   0.000000
37
38    # or plot out differences
39    plot(data6_tukey)
```

Tukey's HSD tests all the pairwise combinations in the data set. In this example, the "sirna2" is different from "sirna1" and "ctrl." It could be that you are not interested in whether "sirna2" is different from "sirna1" and you only care about whether either treatment is different from "ctrl." To answer this question, an alternative is to use Dunnett's post hoc test, specifying "ctrl" as the control. Dunnett's test can be run using the multicomp package in R:

```
40    # Dunnett's test. Use multcomp, install.packages if not
      installed
41    library(multcomp)
42    summary(glht(data6_av, linfct = mcp(rnai = "Dunnett")))
43              Simultaneous Tests for General Linear Hypotheses
44
45    Multiple Comparisons of Means: Dunnett Contrasts
46
47
48    Fit: aov(formula = gfp ~ rnai, data = data6)
49
50    Linear Hypotheses:
51                    Estimate Std. Error   t value   Pr(>|t|)
52    sirna1 - ctrl == 0      1.730      4.618      0.375     0.903
53    sirna2 - ctrl == 0     -98.740     4.618    -21.379    <1e-10   ***
54    - - -
55    Signif. codes:  0 '***' 0.001 '**' 0.01 '*' 0.05 '.' 0.1 ' ' 1
56    (Adjusted p values reported - - single-step method)
```

The nonparametric equivalent to ANOVA is the Kruskal–Wallis test followed by a multiple comparison test (the Dunn–Holland–Wolfe method). In this example, a cellular response is measured after application of two different chemical compounds compared with a control (vehicle). Let's suppose we know from previous experience measuring this response that the distribution is not normal. Another scenario is that you have not collected enough data to know for sure that your measurements are normally distributed. Using nonparametric methods is appropriate in this case, but note that they are more conservative and therefore less powerful. Again using R, we can use the `kruskal.test()` function:

```
1     data7 <- data.frame(
2         response = c(6.0, 9.7, 5.8,
3            6.0, 8.8, 6.5,
4            5.7, 8.5, 5.9,
5            5.6, 9.2, 6.0,
6            5.9, 8.5, 5.6,
7            5.9, 9.0, 6.1,
8            6.3, 9.1, 5.9,
9            6.0, 9.3, 5.0,
10           6.5, 9.0, 6.0,
11           5.7, 9.3, 6.3),
```

```
12    treatment = factor(rep(c("vehicle", "cpd1", "cpd2"), 10))))
13    # carry out the test
14    kruskal.test(response ~ treatment, data7)
15
16        Kruskal-Wallis rank sum test
17
18    data: response by treatment
19    Kruskal-Wallis chi-squared = 19.516, df = 2, p-value =
20    5.783e-05
```

The *p*-value indicates a difference among the three groups. Now the Dunn–Holland–Wolfe test (dunn.test()) will reveal which groups are different:

```
21    # load dunn.test, use install.packages if not installed
22    library(dunn.test)
23    dunn.test(data7$response, data7$treatment)
24        Kruskal-Wallis rank sum test
25
26    data: x and group
27    Kruskal-Wallis chi-squared = 19.5161, df = 2, p-value = 0
28
29
30                    Comparison of x by group
31                        (No adjustment)
32    Col Mean-|
33    Row Mean |      cpd1          cpd2
34    - - - - - - +- - - - - - - - - - - - - - -
35        cpd2 |  3.813029
36             |  0.0001*
37             |
38    vehicle |  3.838534     0.025505
39             |  0.0001*      0.4898
40
41    alpha = 0.05
42    Reject Ho if p <= alpha/2
```

We can see that the responses to compound 1 are different from those of the control (vehicle) or to compound 2.

More Complicated Experimental Designs

Most experiments measure only one thing (designated the *response* variable), and most often they are designed to test the effect of changing one thing at a time (the *explanatory* variable). More complicated experiments can be planned in which more than one explanatory variable is used. For example, we might have a control and three knockout cell lines and want to measure their responses to a drug (two different concentrations plus a control). Because a single quantitative response variable is measured, *univariate* models can be used. The explanatory variables are categorical (the cell lines and the drug treatments are fixed) and so a two-way ANOVA model is appropriate to analyze the data. This will test whether there is an effect that depends on either of the two explanatory variables independently *and* whether there is any interaction between them.

Other designs may include paired measurements on many groups or multiple measurements in two groups. These designs can be analyzed by using repeated-measures ANOVA. These scenarios are rare in cell biology. They are more common in animal experiments in which the same animal receives different concentrations of drugs on different days.

Problems with Data

Data in real life do not always follow textbook examples. You may need to transform them to make more sense and/or perform the statistical tests described above. A number of transformations are possible: scaling, normalization, reexpression, and standardization.

Data can be scaled linearly, $f(x) = a + bx$, so that x is on a more meaningful scale—for example, scaling pixels to real-world values, such as micrometers.

Data might need to be scaled according to the minima and maxima of the data set or of the dynamic range of the instrument for recording, $f(x) = (x - x_{min})/ (x_{max} - x_{min})$. These values are then in the set [0, 1], but can then be scaled to [0, 100] if it makes more sense to express the data as a percentage. Scaling data in this way is important for further operations, such as principal component analyses, that require different data sets to be on a comparable scale.

A few types of normalization are possible. The data might need to be normalized to an independent internal measure to allow a univariate model to apply to the data. For example, the number of instances of a certain feature, such as spots per cell, might be measured. The number of spots may increase as a function of cell size, and the experimental treatment may affect cell size. The number of spots in a cell can be divided by the area of that cell in order to give a comparable measure.

Another example is a time series experiment. Here, fluorescence might be monitored while a drug is applied to the cell. The fluorescence measurements can be normalized to a baseline value before the drug is added to give a measure F/F_0.

Variability between experiments might produce data that need to be normalized to the control values for each experiment in order to make comparisons across the experiments. To do this, each data point should be divided by the mean of the control values. The control values would therefore have a mean of 1 for every experiment, and the values in the other experimental groups would represent a fold change from the control.

One thing to watch out for when normalizing data is that you do not accidentally create a *spurious correlation*. Consider an experiment in which three measurements from each cell are taken, x, y, and z. There are two fluorescence measurements, x and y; z is a measurement of cell size. Imagine x and y show no correlation at all; they are two completely independent signals. Yet, if we normalize to z and rerun the correlation, now all of a sudden they do correlate! This makes sense if you think about it.

Dividing x and y by z puts x and y onto the same scale. Big values of z make small values of both x and y, and vice versa, which creates an artificial correlation.

If the distribution of the data is heavily skewed, the values can be transformed using logarithms to convert them to a normal distribution. This means that parametric statistical approaches can now be applied to the data, which is "reexpressed" on a log scale. Tukey advocated a ladder of transformations to reexpress a data set and checking for skewness as you ascend or descend the ladder.[26]

Finally data can be standardized, $f(x) = (x - \mu)/\sigma$ to give a standard score of the data. Termed a z-score, this transformation is related to the effect size calculation (see below) and to the t-statistic used in the Student's t-test. It is very useful when there are many measurements to compare across different experiments—for example, a large-scale screen or a proteomics data set.

What about individual data points that "look wrong" or skew the analysis? A raw plot of the data you have collected will help to determine if there are "outliers." If you see one (or a few) data points that are far away from the others, they could be outliers. The first thing to check is whether there has been a mistake in the measurement or recording of the data. If there is no obvious problem that can be identified and corrected, next carefully consider removing the point. In most situations, it is best not to remove any data points from the experiment. When planning the experiment, you can set out rules about removal of extraneous points, but it is not a good idea to invent new rules after you have collected data. Two common methods for dealing with outliers are to remove them if they are larger than $Q3 + 1.5 \times IQR$ or smaller than $Q1 - 1.5 \times IQR$. There are other techniques, such as winsorization, but these should be used with caution. By definition, erroneous data points are few and their effect can be minimized by collecting more data.

THE *p*-VALUE

What It Really Means

In the statistical tests above, *p*-values have been mentioned, but what do *p*-values really mean? To the uninitiated, definitions of the *p*-value sound like riddles. There is no way around it: The exact definition and wording is very important. Here is my version. The *p*-value is the probability that you would record similar or more extreme data *if the null hypothesis were true*. Let's say you have measured two groups (control and treatment). The treatment group measurements are much larger than those of control. If there is actually no difference between control and treatment, what is the probability that this difference (or a more extreme difference) will arise again? Note that the *p*-value depends on the variation in the measurements, as well as on the difference between the two groups. If the data vary a lot, a large difference between control and treatment is not unlikely. Whereas if the data only vary a little bit, even small differences might be unlikely.

You might hear other definitions for the *p*-value:

- Small *p*-values mean a large effect, or
- *p*-values tell you the chance that the null hypothesis is false.

These (and others) are incorrect. If you do not understand why, or you think that these subtleties are unimportant, the best thing you can do is to read the article by Steven Goodman that covers 12 common *p*-value misconceptions.[27] You will use statistical tests again and again, and it is important that you understand what the tests tell you, as well as what they do not.

As well as being hard to define, and difficult to understand, *p*-values are controversial. Ronald Fisher, the inventor of the *p*-value, suggested an arbitrary level (<0.05) to distinguish "significance."

However, it is important to note that this is not a magical value at which your experimental results are indisputably correct at $p = 0.0499$ and worthless at $p = 0.0501$. In fact at $p = 0.05$, the false discovery rate (FDR) is at least 29%.[28,29] This has led to a call to redefine the level to 0.005.[30]

Statistically Significant versus Biologically Significant

In cell biology we are concerned with biological significance rather than statistical significance. The reason is that a very small *p*-value can be obtained—giving "statistical significance"—but the difference itself may not actually be *biologically* significant. This phenomenon can occur when very large samples are used for statistical tests.

Imagine the following scenario. You receive a new drug from a collaborator that changes membrane lipid composition. You would like to test whether this change in lipid composition alters the size of the nucleus. You apply the drug and measure the nuclear diameter using microscopy and compare the results to those from cells treated with an inactive compound. You measure a mean of 5.011 μm for the control and 5.019 μm for the compound-treated cells. With an *n* of 3, the *p*-value is >0.05. You now repeat the experiment many more times. Now the means are the same but your *p*-value is 0.0002. This difference is statistically significant! Congratulations! But what does it mean, biologically? On average the nucleus is 8 nm larger in diameter. The percentage change is <0.2%. This is unlikely to be significant in a biological sense.

The take-home message here is that, although we have a convention in biology that $p < 0.05$ means that there "was an effect" in an experiment, we should not rely solely on *p*-values to interpret our data.

This all makes sense, doesn't it? We should not rely on *p*-values; they are only there to tell us how likely the effect we observed was not due to chance. But this leaves us with a problem: How do we assess biological significance?

In the example above, which was willfully extreme in order to make a point, we can see that an 8-nm change in nuclear diameter is nothing to write home about; but what change would be? This is a tough question, and there are no easy answers. Help is on hand in the form of effect sizes and power analysis, but ultimately experimental evidence is required to be able to judge biological significance.

Effect Size

In many experiments, we know that we will get an effect and that that effect will be significant. Here, *p*-values are not very informative. What we really want to know is: What is the size of the effect? This is called *parameter estimation*, where we estimate the size of the effect and the uncertainty around that estimation. A popular method for two samples is to use Cohen's *d* to find the effect size. If the two samples are normally distributed and have similar variance, then the effect size is simply the mean of the test sample minus the mean of the control divided by the pooled standard deviation. Guideline values for Cohen's *d* are that 0.01 is very small, 0.2 is small, 0.5 is medium, 0.8 is large, 1.2 is very large, and 2.0 is huge.[31] These values are only a guideline, because the interpretation of *d* depends on the quality of experimental results and prior experience with similar results.

There are other approaches, such as reporting the difference between the means and a 95% confidence interval. A bias-corrected accelerated (BCa) confidence interval calculated using a bootstrapping[**] approach is a sound method to do this.[32] Because the emphasis with these approaches (termed estimation statistics) is on the magnitude of differences rather than on p-values, they have been referred to as the "New Statistics."[33]

◆ *Golden Rules*

- Good experimental design is key.
- Use the correct statistical procedure for your data.
- Always report n and full details of any statistical tests used.
- Do not lose sight of your scientific question. Your goal is to answer the question and not to achieve a certain statistic.
- Do not rely too heavily on p-values; be open to new approaches.

[**] Bootstrapping is a computational method in which values from the data set are repeatedly sampled randomly with replacement. It can be used to generate BCa confidence intervals. The skewness and bias of the resampled data are corrected during the calculation.

CHAPTER 6

Coding

WHERE TO START

The best advice about how to start coding is *just get going!*

You could decide to enroll in a course to learn coding so that you get a firm understanding of all the concepts. Perhaps you pick up a book from the library to help you. You might just copy and paste from other people's code and work it out from there. There are so many sources to learn how to program computers that there are really no excuses. Whatever works for you, just do it and get going.

The best driver for getting started, whatever approach you take, is having a real problem that you need to solve. The next experiment you do, why not think about how you could use coding to help with the analysis or at least to crunch the numbers that you get from a manual analysis?

Basic Principles: Workflow, Reproducibility, and Benefits

As described earlier in the book, the aim is to write code to do reproducible analysis. This means that:

- Raw data should be read in and processed and outputs generated.
- Outputs should be disposable so that the analysis can be rerun at any time and the outputs regenerated.
- Manual steps should be minimized.
- Ideally, the code should run within one software package. If this is not possible, other packages can be used with automation between them.

Design reproducibility in right from the start—it's best practice. Even if you do not intend to share your code with anyone else, you can reproduce and extend your own work in the future.

Unless the analysis is absolutely trivial, it is always preferable to program a solution. Will it be time-saving? Yes, it will probably take you much longer to program something, especially at the start. However, depending on the time taken to run the manual version and the frequency with which the analysis is performed, it is likely that a programmable solution will ultimately save you time. Even if this is not the case, there are benefits to coding a solution anyway:

1. Robustness: Automation means reducing the chance that a human (you!) makes an error. This is especially true if the analysis is complex, requires more than one or two steps, or is highly repetitive.

2. Reproducibility: You can rerun your analysis to double-check the result. Maybe a data point was incorrectly recorded. Once corrected, the analysis can be run again.

3. Reusability: Someone else can rerun your code on their own data and maybe modify it to add new functionality. That person could be you with your next data set!

4. Auditability: If a manual analysis is run and an error creeps in, it could be difficult to find. By contrast, code is auditable.

5. Learning: Coding is about getting into the habit, and this means the more you code, the better you get and the more you learn. This also means that future code you write will be better, faster, easier to read, and so on

I have noticed that scientists who are used to lab work tend to want to do analysis manually rather than do programming. My suspicion is that this is because they are not fazed by the monotony of repetitive tasks and that they have been conditioned that hard work will crack most problems. Do not be fooled into thinking that manual analysis is faster or that programming will slow you down too much. Manual analysis is unlikely to be faster in the long term, and it is definitely error-prone. Programming is about reducing errors, ensuring reproducibility, and saving you time. It is about making your science go faster.

Mastering the Command Line

Unless you have a background in programming or are already computer-savvy, you might need to overcome "fear" of the command line. Years of using computers with the help of GUIs (graphical user interfaces) mean that some scientists may have developed a fear of interacting with the computer via a text prompt.

There are two reasons why you need to master the command line. First, GUIs will only take you so far. They only present a tiny fraction of what computers can do. Just learning a few text commands can open up huge possibilities for you and your research. Second, programming computers to analyze your data means typing out commands, and the command line is a good place to start. Yes, it is possible to cause a disaster on your computer via the command line, but as long as you understand what you are doing and apply some common sense, it will be okay. The computer tends to ask if you are sure before doing anything too drastic anyway!

Open the Terminal app on a Mac or the equivalent application on Linux. The command line you see is known as the shell and we are using Bash, a Unix shell. Note that it is possible to use these commands on computers running Windows by using cygwin or by installing Linux subsystem for Windows. Try:

```
1    echo 'Hello World!'
```

This command will print Hello World! on the next line. Typing that whole command out again would be annoying; luckily we can just press the up cursor and the command magically appears, ready to be executed again when you press Enter. You can see what commands you have executed previously by using `history`. Previous commands are numbered, and so if you want to run command 342 again, you can do that using !342.

By default, you should currently be in your home directory. Your home directory sits in `Users`, which is in the `root` directory. On the command line, before the prompt, it should state in which directory you are at any point. To change directories, use `cd`. For example, to go to Desktop, use `cd Desktop`. This works because Desktop is a directory in your home directory. Now we are in Desktop. We cannot go to Downloads (for example) by typing `cd Downloads`, because this command means "change to the directory called Downloads within Desktop" and there probably isn't one. However, we can get to Downloads from here in a few different ways:

- Using relative paths: Go up one directory (to home) and then access Downloads. `cd ..` then `cd Downloads`, or by doing `cd ../Downloads`. The current directory is `./`, the one above `../`, and above that, `../../`.

- Using absolute paths: Change directory using the full path—for example, `/Users/YourName/Downloads`. In Bash you can use "tilde expansion" to substitute your home directory in place of the tilde character (i.e., `cd ~/Downloads`).

If you are not sure where you are at any time, type `pwd` to find the present working directory. To get home at anytime, type `cd ~`. To go to root, type `cd /`.

To list contents of the current directory, use `ls`. Give it a try. Almost all commands on the command line can be customized using **flags**. Flags follow a hyphen or double hyphen and change the default settings for a command. For example, `ls` will show visible files and directories by default. To show all files, including hidden ones, use `ls -a`. Flags can be combined after the hyphen so `ls -ltr` is the same as `ls -l -t -r`.

Let's make a directory and manipulate some files. Execute the following lines in your terminal program. The lines that start with a # are comments. Commented lines are not executed and are a great way to explain things to anyone reading your code.

```
1   # go to the Desktop
2   cd ~/Desktop
3   # make a directory
4   mkdir my_folder
5   # is it there?
6   ls -d */
7   # go to my_folder
8   cd my_folder
9   # to see what you are doing, display my_folder in Finder or
    equivalent
10  # make two text files
11  echo "Is this the real life?" > a.txt
12  echo "Is this just fantasy?" > b.txt
13  # check that you have two files called a.txt and b.txt
14  ls
15  # join them together
16  cat a.txt b.txt > c.txt
17  # have a look at the file, and then use q to quit
18  less c.txt
19  # editing the file is possible using nano, use ctrl+x to quit
20  nano c.txt
21  # copy this file out of our folder to the desktop and change its name
22  cp c.txt ../d.txt
23  # now move the file from the desktop into our folder
24  mv ../d.txt d.txt
25  # we can clean up by removing a file with rm
26  rm a.txt
27  # or we can remove all files and then delete our folder
28  cd ~/Desktop
29  rm my_folder/*
30  rmdir my_folder/
```

Asterisks are useful—they mean "anything" most of the time. So `*.tif` can mean `1.tif` or `reallyLongName.tif`. Note how they were used in the `ls` and rm lines above. Because asterisks, spaces, and certain other characters have special meaning, they need to be **escaped** if the command needs the literal character. For example, a folder with a space in the name (my folder) causes problems for cd my folder; the space can be escaped using a backslash cd my\ folder or using literal quotes cd 'my folder'.

The pipe character (|) is very useful on the command line. It allows the output of one program to go directly to another. For example, instead of reading back through the whole of your shell history, you could just pull out the lines you are interested in, using the program grep:

```
1   history | grep echo
```

With this command, the entire output of history is not dumped straight into the terminal; instead it is **piped** to a program called grep. This program can identify lines of text that contain a string—in this case echo. Now, if you ran the examples in this section or ran a command featuring echo at any point in the past, it will be printed in the terminal. Pipes allow chains of programs to link together to do some really powerful things.

The pipe example above was a quick way to find commands that had previously been run. Another useful way to search the history is to use *ctrl+r*. Some other useful tricks include:

- *Tab* can be used to autocomplete. This is very useful when you have long filenames to type out.
- *ctrl+w* deletes to the last space.
- *ctrl+k* deletes everything after the cursor.
- *alt+click* places the cursor (on a Mac).
- *ctrl+c* kills a running command. Useful if you have made a mistake!
- Dragging a folder or file from Finder into the terminal copies its absolute path.

If you are a Mac user you will be used to using the command key for important shortcuts—for example, copy, cut, paste. In the shell, ctrl is a more powerful key. If you are prompted to do ^x—for example, to exit nano—this means *ctrl+x*.

Here are several examples of useful things you can do from the command line.

Recursively copy all files from your Documents directory to a folder on your share on the server:

```
1   rsync -trv ~/Documents/ your_share/Backup
```

Contents of Documents including all subfolders are copied into a directory called Backup in your share on the server. Time stamps (-t) are preserved, copying is recursive (-r), folders and subfolders are copied across, and output is verbose (-v). This is a useful command to run periodically. Anything you have deleted from your computer will stay on the server and any new or updated files will be copied over. Note the slash at the end of the source directory. This means "copy the contents of this directory" and not "copy the directory and its contents."

Are the contents of two folders the same?

```
1    diff -rq ~/Documents/ your_share/Backup/
```

The flags specify a recursive and quiet (no printing of any differences found) comparison of the contents of the two folders. It can be useful to exclude Mac-specific hidden files that are not of interest—for example, ` --exclude='*.DS_Store'`.

Recursive search for files with the tif extension:

```
1    find . -name '*.tif' -print
```

Cannot find a file? Too many folders within folders? Make a file showing all the files and folders in a directory:

```
1    ls -F -R>list.txt
```

Find all ImageJ macro files that feature a method. This is useful if you can remember the method's name but not how you used it or in what macro!

```
1    grep -rnli ~/Documents/my_macros/ -e ij.measure
```

Getting Started with Coding

As discussed earlier, the focus in this book is on Fiji/ImageJ and RStudio/R. The languages for writing ImageJ macros and programming R are easy to get to grips with, and many features of these languages are common to others. Understanding what are variables, strings, arrays, operators, and so on will help you to learn other languages down the line. This will allow you to jump more easily to other platforms or languages later if you wish. The different features may have slightly different names, but just like learning a new spoken language, your understanding can improve by just thinking about similarities and differences between languages as you get the "rules" straight in your own mind.

Once you have mastered the command line and can code in ImageJ and R, you might want to try another language. Python is a powerful high-level programming language that can be used for scripting and numerical analysis—it is possible to control ImageJ from Python. Alternative languages that are useful to cell biologists are C++, Perl, MATLAB, and IGOR Pro.

Only the basics are covered here to get you started with coding. There are a huge number of guides online to help you to become an expert programmer: Seek them out.

Variables and Strings

A **variable** is a thing that can change. This thing can be **numeric, text,** or **Boolean** (true/false). Text variables are commonly referred to as **strings**. Variables are given a name that means that they can be referred to. Variables allow computer programs to be dynamic; when a variable changes, the program runs with it and does not crash. Giving a value to a named variable is known as **assignment**.

A simple example in ImageJ macro language:

```
1  xvar = 2;
2  yvar = 4;
3  result = (xvar + yvar) * yvar;
4  msg = "The result is ";
5  print(msg, result);
```

And in R:

```
1  xvar <- 2
2  yvar = 4 # you can use <- or = for assignments
3  result <- (xvar + yvar) * yvar
4  msg <- "The result is "
5  print(msg)
6  print(result) # can be printed together if required
```

In both cases xvar and yvar are numeric variables, whereas msg is a string. Variables can change, but those examples were not dynamic. Instead, the variables could be assigned values that are retrieved from an image, for example. We do not know what those values are beforehand, so they must be assigned by a program *dynamically*. For example, in R, run:

```
1   xvar <- runif(1,0,100)
2   xvar
```

Now run the code again. Note that the value of xvar changes. We do not know before running this example what value xvar will take (we know it will be between 0 and 100). It is assigned dynamically as the program runs.

Arrays and Vectors

Collections of variables are referred to as arrays in ImageJ and vectors in R. They are useful because the contents can be set dynamically, and because they are ordered by an index, each element can be worked with by referring to its index position. In R there are other types of data structures, including lists, matrices, and data frames. All of these data structures are very useful when writing code:

```
1   # simple example of vectors
2   x <- c(3, 5, 1, 9, 0)
3   # print the 3rd element
4   x[3]
5   # now assign some random values instead
6   x <- rnorm(10,10,2)
7   # sum all except the first item
8   sum(x[-1])
9   # run the last two commands again, note the change in value
```

Note that when we printed the third element, we used x[3]. This is because R uses 1-based indexing, not 0-based. Many languages count from 0 and this difference causes problems when switching between languages. ImageJ is essentially 0-based (as we will see next), but note that image stacks and hyperstacks break this convention. Pixels' x and y locations are indexed from 0, as are their values, but other dimensions (C, Z, and T) are 1-based!

Loops

Loops are wonderful things. They allow a little block of code to run again and again, until it completes and the program can move on. Loops are very useful for working on arrays, vectors, and lists. In ImageJ macro language there are three types:

1. `for`—The code runs a predefined number of times (e.g., once for every image in a directory).

2. `while`—The code runs while a certain condition is true.

3. `do-while`—The code runs once and then will run repeatedly while a certain condition is true.

Examples of these three loops in ImageJ:

The code to be run is enclosed in brackets. The `for` statement has three components separated by semicolons—initialize; condition; increment:

```
1     // example of for loop
2     for (i=0; i<10; i++) {
3            j = 5*i;
4            print(j);
5            }
```

This loop initializes with the variable `i` set at 0. It runs the code and then 1 is added to `i` (`i++` means i+1); the condition is tested, and if it is true, the code runs again. So the second time through, `i=1`, and because this is <10, the code will run. This code will run print 0,5,10,15,...,45.

The next loop type is a `while` loop, which will give the same result as the `for` loop above. The condition is tested before the loop is run, as above:

```
6     // example of while loop
7     i=0;
8     while (i<=90) {
9            print(i);
10           i = i + 5;
11           }
```

A `do-while` loop tests the condition at the end of the loop rather than at the start. This means that, whatever the condition, the code will run once:

```
12    // example of do-while loop
13    i=0;
14    do {
15           print(i);
16           i = i + 5;
17           } while (i<=90);
```

With a `for` loop, you generally need to know how many loops will run when you start the code. Using `while` or `do-while` loops is more open-ended. They are ideal when you are not sure how many times it will run or when the number of times changes depending on what happens while the program runs.

In R, there are `for` loops as well as `while` and `repeat` loops. An equivalent `for` loop is shown here:

```
1    # example of for loop
2    j <- 0
3    for(i in 1:10) {
4            j = 5*(i-1)
5            print(j)
6            }
7    # note that the apply family of R functions mean that loops are not
     always needed
```

Because loops can run many times, a general rule is to keep the lines of code in the loop to a minimum. Anything that can go outside the loop should be put outside the loop, so that it does not slow your code down.

Loops can be coupled with conditional statements to further increase their power. For example, you could run a loop `if` a certain condition is met. Or you could do something different if the condition is anything `else`. The `if` statement could be within a loop to make the loop more selective:

```
18    // adding complexity to a for loop
19    for (i=0; i<10; i++) {
20            if (i%2==0) {
21                j = 5*i;
22                print(j);
23                }
24            }
```

This loop will now only print 0,10,...,40 and not 0,5,...,45. The `if` statement can be extended with an `else` condition and/or a number of `elseif` conditions. Other embellishments are possible. For example, a `break` statement can be added to prematurely terminate a loop.

Some next steps to try:

1. Nest a `for` loop within another to create a matrix of numbers.
2. Is there a way to make this matrix in R without writing loops?

In the code examples above, we used **operators** like == and >=. It is worth memorizing the most common operators; they're mostly the same in other languages (but not always!):

Description	Operator	Software
Assignment	<- or =	R
	=	ImageJ
Addition	+	Both
Subtraction	-	Both
Multiplication	*	Both
Division	/	Both
Exponentiation	x^y or x**y	R
	pow(x,y)	ImageJ
Modulo (x mod y)	x %% y	R
	x % y	ImageJ
Less than	<	Both
Less than or equal to	<=	Both
Greater than	>	Both
Greater than or equal to	>=	Both
Equal to	==	Both
Not equal to	!=	Both
x OR y (Boolean)	x || y	Both
x AND y (Boolean)	x && y	Both

HOW TO WRITE A BASIC ImageJ MACRO

Macro languages allow the control of a computer program's main functions without having to point and click at everything repeatedly. This means that operations are limited but, even so, writing an ImageJ macro can save lots of time and is easy to do. A great way to start writing a macro is to use the recorder in Fiji. When in record mode, the recorder puts in a window all the commands that are executed. You can use this as the basis for a new macro without having to look up all the commands. In fact, using the recorder is good practice while you are working in Fiji even if you are not writing any code, because it keeps a "history" of what you've done (like in Bash or RStudio).

With an image open, click *Plugins > Macros > Record…*, and carry out the operations you would like to add to your program. Notice how the commands are added to the window. When you are finished, click "Create" and a new Macro window containing your commands will appear. You can now edit this to make your script.

Working on All Files in a Directory

Say we want to measure the mean intensity of every image file in a folder. The basic macro below shows how to do this.[*] You can use this as a template for your own purposes by replacing the commands within the loop:

```
1    /*
2    * Comments here are useful to describe the macro
3    * This macro will measure the mean of each image in a folder
4    */
5    // Prompt user to select the folder with images
6    dir = getDirectory("Select the source directory");
7    // Make a list of all files in this directory
8    list = getFileList(dir);
9    // Sort the file list alphabetically
10   Array.sort(list);
11   // Use batch mode: speeds up execution.
12   setBatchMode(true);
13   // Define measurements we want
14   run("Set Measurements...", "display area mean redirect=None
     decimal=3");
15   // use a loop to run command(s) on each image
16   for(i=0; i<list.length; i++) {
17        // define the ith file in the list and open it
18        filename = dir + list[i];
19        open(filename);
20        // run these commands on each image
21        run("Select All");
22        run("Measure");
23        run("Close All");
24        }
```

---

[*] Notice how the commenting syntax is different from Bash. Also, note the difference between single-line and multiline commenting.

The three lines in the loop will select, measure, and close the image. You can substitute these three lines with whatever you want. Note that Fiji offers a similar built-in template from the *Templates* menu in the macro window. A basic macro like this can get you quite far, but as it stands it has some problems.

What if the user has a directory that contains nonimage files? This might cause a crash, so we need to deal with this case. Maybe we also want to extend the code to do something more than just measure the images. Let's say we now want to threshold the image and save this modified version in another directory:

```
1   /*
2   * This macro will save a binarized version of each image in a new
    folder
3   */
4   dir1 = getDirectory("Select the source directory");
5   // User must now also pick a destination folder
6   dir2 = getDirectory("Select destination directory");
7   list = getFileList(dir1);
8   Array.sort(list);
9   setBatchMode(true);
10  for(i=0; i<list.length; i++) {
11          filename=dir1+list[i];
12          // check we are dealing with a tiff file
13          if (endsWith(filename, "tif")) {
14                  open(filename);
15                  // this part contains the operations for each image
16                  setAutoThreshold("Otsu");
17                  setOption("BlackBackground", false);
18                  run("Convert to Mask");
19                  // now save the modified image in the destination folder
20                  saveAs("TIFF", dir2+list[i]);
21                  close();
22                  }
23          }
```

A further change is to externalize the operations for each image into a function. This improves readability and helps with editing and extending the macro:

```
13          if (endsWith(filename, "tif")) {
14                  open(filename);
15                  // use a function to do the operations for each image
16                  processImage();
```

```
17                          // now save the modified image in the destination folder
18                          saveAs("TIFF", dir2+list[i]);
19                          close();
20                          }
21          }
22    function processImage() {
23                setAutoThreshold("Otsu");
24                setOption("BlackBackground", false);
25                run("Convert to Mask");
26          }
```

Use these examples as a template that can be modified to get ImageJ to do what you need. Some next steps to try:

1. Change the code in the loop so that it will do something different.
2. Modify dir2 so that the destination folder is specified to save the user a few clicks.
3. Alter the filename of the modified files to show that they have been changed.

▶ **Tutorial: Blinding Files for Manual Image Analysis**

In the Tutorial: Quantifying Cell Protein Levels from Immunofluorescence Images in Chapter 4, we used manual image analysis. To eliminate bias, a user blind to the conditions should do the manual analysis. Even if an independent person does the analysis, they might still be influenced by the filename or by the labeling of the images. This simple macro (pages 104–106) anonymizes all images in a directory. It strips out the labeling from the images and also provides a log file so that the analysis can be matched back to the original files at the end.

```
1     /*
2     * This macro will prepare a directory of TIFFs for blind analysis
3     * and log the association between original file and blind analysis
      file
4     * for unblinding at the end of the analysis
5     */
6     macro "Blind Analysis" {
7             dirPath=getDirectory("Select a directory");
8             // Get all file names
9             allNames=getFileList(dirPath);
10            // Create the output folder
```

```
11        outputDir=dirPath+"blind"+File.separator;
12        File.makeDirectory(outputDir);
13        // Make an array and extend it with names of *.tif only
14        imNames=newArray(0);
15        for (i=0; i<allNames.length; i ++) {
16              if (endsWith(allNames[i], ".tif")) {
17                      imNames=append(imNames, allNames[i]);
18              }
19        }
20        imNum=imNames.length
21        // Generate a permutation array of length imNum
22        imPerm=newArray(imNum);
23        for(i = 0; i < imNum; i ++) {
24            imPerm[i]=i+1;
25        }
26        // Shuffle the array
27        for(i=0; i<imNum; i ++) {
28              j = floor(random * imNum);
29              swap=imPerm[i];
30              imPerm[i]=imPerm[j];
31              imPerm[j]=swap;
32        }
33        // Associate sequentially permuted positions to image names
34        imPermNames=newArray(imNum);
35        for(i=0; i<imNum; i ++) {
36              imPermNames[i]="blind_"+IJ.pad(imPerm[i],4); //
                  example name: "blind_0001"
37        }
38        // Open each image, strip metadata, save in destination
          folder using the blinded name
39        // Also log both names in the log.txt file created in the
          destination folder
40        setBatchMode(true);
41        f=File.open(outputDir+"log.txt");
42        print(f, "Original_Name\tBlinded_Name"); // tab separated
43        for(i=0; i<imNum; i ++) {
44              inputPath=dirPath+imNames[i];
45              outputPathPerm=outputDir+imPermNames[i];
46              open(inputPath);
47              totalSlices=nSlices;
48              if(totalSlices>1) {
```

```
49                         stripFrameByFrame(totalSlices);
50                  } else    {
51                         setMetadata("Label", "");  // strips the label
52                  }
53                  save(outputPathPerm);
54                  print(f,imNames[i]+"\t"+imPermNames[i]);
55                  close();
56          }
57      setBatchMode("exit and display");
58      showStatus("finished");
59  }
60
61  // strips the label data from each slice of an image
62  function stripFrameByFrame(totalSlices) {
63          for(i=0; i<totalSlices; i ++){
64                  setSlice(i+1);
65                  setMetadata("Label", "");
66          }
67  }
68
69  // function that adds a variable to an array
70  function append(arr, value) {
71          arr2 = newArray(arr.length + 1);
72          for (i = 0; i < arr.length; i ++)
73                      arr2[i] = arr[i];
74                      arr2[arr.length] = value;
75          return arr2;
76  }
```

HOW TO WRITE A BASIC R SCRIPT FOR ANALYSIS

Scripts in R are a convenient way to automate analysis. RStudio provides a nice window to write and test your script, running each element line by line. In this example we will import some data, do some analysis, and plot the result.

A user has opened a movie of a cell in Fiji and manually positioned several ROIs in ROI Manager. They executed the multimeasure command and saved the Results as a csv (the default name is "Results.csv"). The goal is to calculate the average value of all ROIs per frame and plot out the data and save the file. This is therefore part of a *workflow*, in which the first part is done in Fiji and the final part in R.

Open RStudio and start a new project; save it in a folder somewhere on your computer. The folder that contains the project will be your working directory. Copy your Results.csv file there:

```
1    # Start with a comment about the purpose of the script
2    # This code will analyze the multimeasure output from ImageJ
3    # import data
4    my_raw_data <- read.csv(file='Results.csv', header=TRUE,
     stringsAsFactors=FALSE)
5    # find the names of all columns
6    my_col_names <- colnames(my_raw_data)
7    # now find the columns named Mean* for each ROI
8    mean_columns <- my_col_names[grepl("^Mean",my_col_names)]
9    # make data frame with only mean columns
10   my_data <- subset(my_raw_data, select=mean_columns)
11   # have a look at each ROI
12   matplot(1:length(my_means), my_data,
13         type = "l",
14         lty = 1,
15         col = "grey",
16         xlab = "Frames",
17         ylab = "Mean Pixel Density")
18   # calculate means per row
19   my_means <- rowMeans(my_data, na.rm=TRUE)
20   # open a pdf file
21   pdf("plot.pdf")
22   # plot out the result
23   plot(1:length(my_means), my_means,
24         type = "l",
25         col = "red",
26         lwd = 3,
27         xlab = "Frames",
28         ylab = "Mean Pixel Density")
29   # close the pdf file
30   dev.off()
31   # save the averaged data
32   write.csv(my_means, file = "output.csv",row.names=FALSE)
```

This basic script will do what we want. It can also deal with variations between data sets that would otherwise cause a problem. For example, it does not matter how many ROIs the user has drawn; the script will find them. Movies can vary in length, and as long as the user has specified that the mean goes to the results window in Fiji, using *Set Measurements...* the script will work. If additional measurements are selected, the script will still work. However, we will soon want to add more functionality.

For example, if we want to analyze many csv files from different movies, the script would need to be run each time. We can edit the script so that all csv files in a directory are analyzed (regardless of their filename). Put the csv files to be analyzed into a folder called data, within the project folder. Now the code needs to make a list of all csv files and then process each one in a loop. We can also simplify the code by specifying a function to do the processing and then just call that function for each file we want to process:

```
1   # Script to process multiple csv files containing output from
    multi-measure
2   #
3   # search all .csv files in current working directory
4   my_files<-list.files("./data/",pattern='*.csv',full.names=T)
5   # make directory for output if it doesn't exist
6   if (dir.exists("output")==FALSE) dir.create("output")
7   # function definition
8   calc_and_plot <- function(my_filename) {
9     # import data
10    my_raw_data <- read.csv(file=my_filename, header=TRUE,
    stringsAsFactors=FALSE)
11    # find the names of all columns
12    my_col_names <- colnames(my_raw_data)
13    # now find the columns named Mean* for each ROI
14    mean_columns <- my_col_names[grepl("^Mean",my_col_names)]
15    # make data frame with only mean columns
16    my_data <- subset(my_raw_data, select=mean_columns)
17    # calculate means per row
18    my_means <- rowMeans(my_data, na.rm=TRUE)
19    # retrieve the filename and extension from full path
20    old_filename <- basename(my_filename)
21    # output files will have "out_*" names
22    my_output <- paste("out_",old_filename, sep="")
23    # convert *.csv name to *.pdf
24    pdf_name <- paste(substr(my_output, 1, nchar(my_output)-4),".
    pdf",sep="")
```

```
25    # create path to output directory
26    out_path <- file.path(getwd(),"output",pdf_name)
27    # open a pdf file
28    pdf(out_path)
29    # plot out the result
30    plot(1:length(my_means), my_means,
31         type="l",
32         col="red",
33         lwd=3,
34         xlab="Frames",
35         ylab="Mean Pixel Density")
36    # close the pdf file
37    dev.off()
38    # save the averaged data
39    out_path <- file.path(getwd(),"output",my_output)
40    write.csv(my_means, file=out_path,row.names=FALSE)
41  }
42  # call the function for each file in the list
43  for(i in 1:length(my_files)){
44    my_filename <- my_files[i]
45    calc_and_plot(my_filename)
46  }
```

In the first example, the names of the outputs (plot.pdf and output.csv) were hard-coded and saved in the same directory. This would cause problems for a script in which multiple input files are processed, because the outputs would be overwritten. In the new function, the output files are named according to the input file. They are also saved into an output directory. If they were saved into the working directory, and the user reran the script, the output csv files would be processed along with the original data. Note how we are processing data from one directory and make outputs in a different directory, according to the principles described at the beginning of the chapter. The outputs of both of these simple scripts are shown in Figure 6.1.

WHAT CAN GO WRONG?

How to Validate and Check Your Data

Skepticism is essential when programming a workflow. Suppose your program "works." You pointed it at a directory of images and after some churning, out popped

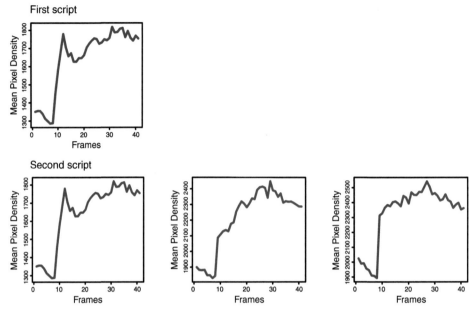

FIGURE 6.1. Outputs from the two R Scripts.

a graph summarizing the analysis. This is a reason to rejoice (your program has worked!), but now you need to check that everything worked as it should. Just because the program ran without crashing does not mean that the analysis itself is correct or that the right thing was plotted. There are many possible errors even in a simple workflow, and the number only increases for longer, more complex programs. Here's the best bit: All that automatic processing that is so nice and time-saving and hidden, it conspires against you, preventing you from discovering errors. So what do you do?

Often a critical eye is all that is needed to spot problems. For example, suppose you only have 11 data points on your graph but 12 images in the directory. This can indicate that you have an off-by-one error in a loop or a file that is not compatible with the program. Sometimes the result just does not match what you can see with your eyes. In these cases it is a good idea to use simulated data for which the result is known (so-called "ground truth" data).

For example, in R, you can quickly generate some randomized data to check a function. To make an $m \times n$ matrix of normally distributed data do the following:

```
1    set.seed(1)
2    m <- 80 # rows
3    n <- 200 # columns
4    # rnorm default is mean=0, sd=1
5    data0 <- replicate(n, rnorm(m))
6    # use data0 e.g. find mean of each column
7    colMeans(data0)
```

If a reproducible set of randomly generated numbers is needed, set.seed(1) can be used, where 1 is an integer, and this matrix of numbers can then be reused in other codes.

For ImageJ, blank or noise images can easily be generated. These can be useful for feeding into a workflow to check that the result is as expected. More sophisticated sets of dummy images can be produced, as described by the code below. Imagine a case where your code should be able to recognize spots in microscope images automatically and perform some analysis. You are not sure how well it performs. The code below will generate a noisy (i.e., realistic) image with 10 spots that are weakly brighter than the background. Importantly it also makes a "ground truth" image that unambiguously shows the true location of spots.

```
1    /*
2    * Make a test image and a ground truth image to compare.
3    * Ten spots on a gray noisy background
4    */
5    setBatchMode(true);
6    imageDim=512;   // xy Dimensions of images in pixels
7    // make an image of noise
8    newImage("TestImg", "8-bit random", imageDim, imageDim, 1);
9    // apply a mean filter
10   run("Mean...", "radius=2");
11   // retrieve the ID of the TestImg window
12   Tid = getImageID();
13   // make a black background ground truth image
14   newImage("GTImg", "8-bit grayscale-mode black", imageDim,
         imageDim, 1);
15   // retrieve the ID of the GTImg window
16   GTid = getImageID();
17   // set foreground color to white
```

```
18   setColor(255);
19   // in a loop, position spots on each image
20   for (i=0;i<10;i++) {
21          // random position in the image
22          posX=floor(random*imageDim);
23          posY=floor(random*imageDim);
24          selectImage(Tid);
25          // fill a 5-pixel circle at this position
26          fillOval(posX, posY, 3, 3);
27          selectImage(GTid);
28          // do the same in ground truth image
29          fillOval(posX, posY, 3, 3);
30   }
31   selectImage(Tid);
32   // blur the whole test image
33   run("Gaussian Blur...", "sigma=2 stack");
34   setBatchMode("exit and display");
```

This simple code can easily be extended. Next steps:

1. Rerun the code making more spots or bigger ones.

2. Modify the code so that it will save the two images.

3. How could you change this code so that it made 100 pairs of images, saving them automatically?

Debugging

Debugging is an essential part of coding. The easiest bugs to squash, but often the most frustrating, are the ones that simply prevent the program running to completion or that cause a crash. They can be as small as a missing semicolon or the presence of a lowercase letter where an uppercase one should be. You will be amazed at how obvious they are when you find them! Once you do find and fix them, your program runs, but now the bugs that remain are hidden. They can sometimes only reveal themselves when other people start using your code. For example, the user might type in "15" and find the program crashes. Of course, as the programmer you know that 15 is a ridiculous number that no one in their right mind would type in. You were wrong, and somebody has. Shoring up your code to prevent all kinds of user behavior is the next level in coding, but it will help build trust in your program if people can run it without it crashing.

There are some good techniques for finding errors and for understanding what has gone wrong:

1. Error messages can be informative; take time to find out what they are trying to tell you.
2. A debugger module may be present in the software specifically to find and fix errors.
3. Print statements can be useful for revealing the state of the program and identifying problems.

Getting Help

When you start out, you might find yourself googling every single line of code you write. Even experienced programmers have to google for solutions to particular problems, so there is no shame in seeking help when learning to program. Websites like Stack Overflow are wonderful resources of previous questions and "best answers" that are upvoted or downvoted on the basis of their usefulness. It is highly likely that whatever problem you are having has already been encountered and solved. Knowing how to quickly search and find solutions is actually a key programming skill. Using the exact error code or message as a search term is a good way to start.

Sometimes you'll find the answer most quickly by reading the manual:

- For command line tools, type `man NameOfPackage` to display the manual of the package you are using.
- In RStudio, you can search help or type `?function` to display help for the function you are interested in.
- In Fiji, there are pages on the ImageJ website that describe the main functions. Plug-ins tend to have their own documentation.

If you still draw a blank and there is no local expertise, it is time to ask a question yourself online. Seek out the best place (where questions get answered quickly): This might be a forum, mailing list, or subreddit. Some quick tips to get the best answers:

- Be very specific in your subject line—for example, do not write "ImageJ Question"; instead try "How do I export a line profile from ImageJ in csv format?"
- For the actual question, keep it succinct, but make sure you include details such as your operating system, the version number of the software, and any plug-ins that you are using.

- It is best to include a MWE (minimal working example). This is ideally just a few lines of code that demonstrate the problem and will make it easier for people to verify it and then help you.
- Include a description of the things you have tried or searches and links you have found and why they have not helped.
- Always be polite and courteous. You might be very frustrated when you post your question, but bear in mind that the people who help are often the authors of the software package and are answering questions as a community service.

Sometimes, simply typing out a question, or making a MWE, solves the problem for you. The act of explaining your problem from scratch can sometimes reveal the problem or the bug in your code. This is actually a legitimate programming technique known as "rubber duck programming," in which the programmer explains the code line by line to a rubber duck (or other inanimate object) until the bug is found. Sounds crazy, but it works, by making you take a step back and see the errors.

GETTING GOOD

Once you have written some code that runs and you begin to reap the benefits of programming, you will begin to get good at it. Some "next level" coding advice follows, but you should keep it in mind right from the start.

Ugly Code

After a while, you will look back at your first few attempts at coding and notice how "ugly" it looks. This is because, over time, you will pick up tricks to make your code easier to read and understand. An excellent way to accelerate this is to look at other people's code to get inspiration. Another is to share your code and allow people to comment on it. Even without feedback, simply getting your code ready to share means that you have to focus on making it understandable to someone else. The result is that this helps *you*, so that you can read and understand your own code in the future.

There are simple things you can do to improve readability. Pay attention to whitespace. Indenting loops, for example, using a tab character or a set number of spaces greatly increases the readability. Note that some languages, such as Python, are sensitive to whitespace, which means that you have to pay attention to it! Another is to comment frequently. When writing your first code, write a comment for each line; it will help you to understand what you are doing. Having a simple system for naming variables can help a lot. Variables like `n1xSz` may make sense to you at the time, but `xDimensionSizeImage1` is more understandable later.

Write Modular Code

An extension of readability is writing so-called modular code. In simple terms this means writing separate routines for small tasks in a bigger program. The main program runs and then when the task is needed, the routine is called; when it finishes, the main program continues. This helps readability, because it can help you to see the logic of the main program without getting bogged down in the details of smaller tasks. The other benefit here is that writing modular code means that you can test whether each module works without worrying about the rest of the program (unit testing). This speeds up the process of correcting bugs in your code. For example, if you know that all modules run correctly, but the main program crashes, then the bug is likely to be in the main program. Finally, because these modules are self-contained, they can be accessed by other programs that you write. This saves you time because when you update the module, all of your main programs will be "updated" without having to go into each and make the same correction.

Version Control and git

Version control is an important part of coding and organizing your work. It is a system of recording all the versions of your program while you are developing it. This means that you can roll back to previous versions if something goes wrong or if the changes you make do not work out for some reason. Behind the scenes, it all works by comparing the files line by line using dedicated software, usually **git**. The different versions are stored in a database called a **repository** (**repo**). You can run all of this locally on your computer and keep it in sync with a remote website such as GitHub or Bitbucket.

Version control is similar to having a super Undo button, but instead of going back change by change to the last saved version, you can skip back through saved versions and easily find which version you need. This means you need to save each version of your code that you might want to go back to. In git this is called a **commit**, where you write a short commit message to explain the state of the code or the changes you have made. Each commit has a specific hash, a unique string that acts as a referrer. When writing up your notebook or your research paper, you can refer to the hash that was used for your analysis. This means that any future developments can happen and readers can still access the version of your code that was used. An alternative method for complete projects is to obtain a digital

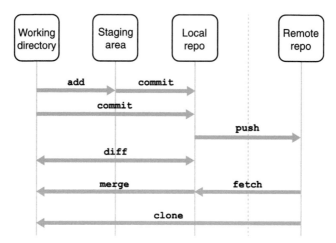

FIGURE 6.2. How git works.
Schematic diagram to show how version control using git works. See the text for details. Note that the **pull** command is a combination of fetch and merge, to ensure you have the latest version of the repo. **clone** is a way of obtaining the entire repo for the first time.

object identifier (DOI) for your code so that it can be cited by others. You can do this by pushing your repo to Zenodo, which will issue a DOI.

Figure 6.2 shows how version control using git works. The bit of code you are working on is in a *working directory* on your computer. To allow git to start watching this file, you need to **add** it. The file is now in the *staging area* but is not yet in your *local repo*. To put it into the repo you need to **commit** it. If you are working on your own on this coding project and you only have one computer, this is all you need to work locally. You can compare the subsequent changes you make using **diff**.[†] If you are happy with those changes, you can do a new commit; if you need to revert to the last commit, you can use the **checkout** command. The nice thing about version control is that you can **push** your code to a *remote repo* (such as GitHub). This means your code is safe if you run into a problem and it also means you can easily work on the code on another computer. To do this you **clone** the whole repo to make a local repo on another computer and get going. From then on you can **fetch** the latest changes to each local repo. Everything stays in sync and you are always working on the latest version of your code. You can use git on the command line using the terminology described above or you can use a GUI such as Sourcetree or GitHub Desktop. RStudio incorporates version control using git on a project.

[†] diff is a command line tool that compares two files and finds any lines that differ in content.

The other benefit of using a remote repo is that it allows other people to work with you on your code. Two other bits of terminology you will encounter are **fork** and **merge**. Forking is to make a new **branch** of the code so that you can try out new things. If this goes horribly wrong, or it was just a bad idea, you can kill the branch with no consequence to the main code, usually the **master branch**. If your branch was a success you can merge it back to the master—as long as it has not been changed in a way that makes this impossible. If there have been changes to the main branch, you will need to resolve these conflicts before you can merge.

Version control is used by advanced programmers; however, it is even more useful to people getting started with coding, because rolling back changes after making mistakes is very valuable. Unfortunately, understanding version control is yet one more thing to learn at a time when you are trying to learn to write code. One useful suggestion is after you have completed your first program and it is running well, **commit** your first version to a repo and get in the habit of using version control from this point on. It will save you a lot of time in the long run. An alternative suggestion is that if you are starting out using someone else's code in a public repository, do not download the code and start to work on it, instead create an account and **fork** their code and begin work in this repository.

Sharing Your Code

Once you have some useful code, you will probably want to share it with other people in your lab. Of course you can send them a copy of the code, but what if you update it? Then you'll need to send them a new version. What about sharing it with more people or with the rest of the world? You need a way to make the latest version of your code available to anyone who is interested. One advantage of a remote repo such as GitHub is that it can be public. The latest version (and any previous versions) are available for anybody who wants them. Interested people can *watch* the repo and get alerts when a new commit is pushed.

Fiji has a feature called ImageJ Update Sites. This allows people to subscribe to your Update Site so that they can use your code in Fiji. Importantly, any changes to your code are pushed to the subscriber without them having to make any effort. If you only want to share your code with a few people and do not want the world to see it, there are a few solutions. You can put your code on a lab server share that is accessible to all members. Alternatively, you can work on a remote repo that is private and allow access only to certain users in your organization. Ultimately, when you publish your work, you will need to make your code available. So working with a repo that can be made public easily is the best solution. For academics,

GitHub offers the ability to have private remote repos and make them public at a later date.

◆ *Golden Rules*

- Commenting is essential to help the future you and anyone else who will use your code.
- Readability is very helpful. Take some time to format your code nicely.
- Try to write generic code that can be easily reused.
- Pick meaningful, easy-to-understand names for objects, variables, functions, etc.
- Use a consistent style for your code.
- Externalize any small contained blocks of code into functions that can be accessed by different scripts and macros.
- When you begin to code, the learning curve is steep. Don't be afraid to ask for help.
- Design reproducibility into your workflow or pipeline right from the start.
- Do not delay starting to code: Just get going!

CHAPTER 7

Putting It Together

PLOTTING DATA

There are many different ways to display quantitative information. Around 40 graph types exist—many more if variations on these basic types are considered. In cell biology, only a fraction of these graph types are useful. Figure 7.1 shows the most common graph types and their main purposes.

Best Practice

Which graph type should we use and when? Let's focus on the presentation of experimental results in which a variable has been measured. There is one measure for each experimental unit (n), and there may be a number of experimental groups (Table 7.1). We want to display the *distribution* of the data. Single measurements can be described in the text or a table, as can summary statistics for experiments with low n and/or few experimental groups. Once n becomes large, it is best to show the results in a histogram for a single experimental group. If there are more experimental groups, histograms cannot be easily overlaid, and so scatter, box, and violin plots should be used. Scatter plots work well for lower n and violins work well for large n (>100). Bar charts should only be used for single measurements, to compare several experimental groups. When many groups are compared, bar charts are easier to interpret with the categories arranged vertically and the bars extending horizontally (to do this, use `coordflip()` in R). Note that bar charts should not be used for the display of summary statistics alone (i.e., average ± error); it is always best to show the underlying data.[34]

There are whole books dedicated to the art of data visualization and presentation of scientific results.[35] Highlighted below are some common themes that emerge, together with a number of established principles for scientific data presentation that you should adhere to.[36]

- "Chart junk" refers to all excess lines and notation that do not contribute to the main message of the graph. The aim is to have a high data-to-ink ratio. Generally, if you delete chart junk, the graph will improve.

Aim	Plot	Names	Notes

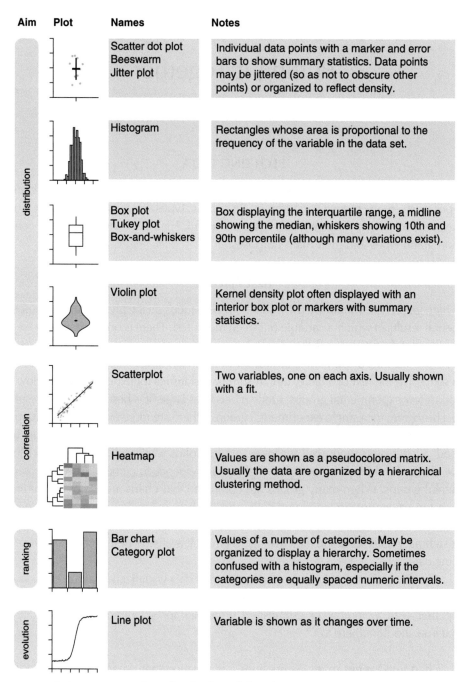

	Scatter dot plot Beeswarm Jitter plot	Individual data points with a marker and error bars to show summary statistics. Data points may be jittered (so as not to obscure other points) or organized to reflect density.
distribution	Histogram	Rectangles whose area is proportional to the frequency of the variable in the data set.
	Box plot Tukey plot Box-and-whiskers	Box displaying the interquartile range, a midline showing the median, whiskers showing 10th and 90th percentile (although many variations exist).
	Violin plot	Kernel density plot often displayed with an interior box plot or markers with summary statistics.
correlation	Scatterplot	Two variables, one on each axis. Usually shown with a fit.
	Heatmap	Values are shown as a pseudocolored matrix. Usually the data are organized by a hierarchical clustering method.
ranking	Bar chart Category plot	Values of a number of categories. May be organized to display a hierarchy. Sometimes confused with a histogram, especially if the categories are equally spaced numeric intervals.
evolution	Line plot	Variable is shown as it changes over time.

FIGURE 7.1. Options for display of data from a typical experiment.
The most common graph types used in cell biology.

TABLE 7.1. *Plot type depends on the number of observations and the number of experimental groups*

	Experimental groups			
n	1	1–3	4+	10+
1	Text	Text Table Bar	Table Bar	Bar
2–10	Text	Text Table Scatter	Table Scatter	Scatter
11–99	Histogram	Histogram Scatter Box	Scatter Box	Scatter Box
100+	Histogram	Histogram Box Violin	Box Violin	Box Violin

- Avoid 3D representation if a 2D plot will do the same job. For example, 3D bar charts offer nothing over standard 2D plots. They are much harder to read and should be avoided.
- Axes should start at the origin. If not, a break should be indicated. An exception to this rule is $\log_{10}$ axes, where it is impossible to include 0.
- The *y*-axis should show data that depends on the variable on the *x*-axis. In an experiment, the conditions you control go on the *x*-axis, whereas your measurement is plotted on the *y*-axis.
- If the data are calculated as a ratio it is usually best to present them on a log scale. On a $\log_2$ scale, a twofold increase (2, 1) or decrease (0.5, −1) is symmetrical about 1 (0 on a $\log_2$ scale).
- Logarithmic presentation on the *y*-axis makes sense for data types that vary widely (e.g., proteomic data, exponential cell growth). Logarithmic presentation on the *x*-axis makes sense for a range of concentrations (e.g., concentration–effect plot).
- When making more than one plot of the same kind of data, use consistent ranges on the axes. This makes comparisons between plots easier.
- When plotting more than one data set on the same plot, use markers that can be easily distinguished (color, shape, size). Patterns and new interpretations may emerge from such plots, but consider plotting separately if there is significant overlap.

- If every data point is plotted, none should be hidden. Do not cover any points with labels. If points are very dense, use transparency, smaller markers, or another "overplotting" technique to make sure the data are not obscured.

- Color use is effective but can be overused. Be aware of color blindness when picking color palettes. Cynthia Brewer's ColorBrewer is an excellent resource.

- Avoid "rainbow" palettes such as jet—our eyes do not perceive the gradient of colors evenly. Try to use instead a perceptually uniform color lookup table such as mpl-viridis.

- Be wary of using graph types favored by mainstream media such as pie charts. There is almost always a better way to show the data.

MAKING FIGURES

It is best to get into the habit of making figures early. Without assembling figures or plotting out your data, it is difficult to know if your experimental work or analyses are complete. We still need to show representative images in our papers, and these must reflect the graphs shown in the same figure to support the result. Sometimes an experiment has been repeated many times, and yet the images might not be "nice enough" for presentation. Alternatively, you might have overlooked an important control and need to go back and repeat the experiment again. Assembling a figure is the only way to check if you are "over the line" and are truly finished with that set of experiments. Start as soon as possible; do not wait until you are writing the paper to assemble the figures.

These days, a figure in a paper has multiple panels. Each panel may have one (or more) plots or a montage of images. Therefore each figure panel needs to be generated separately and then the figure as a whole assembled.

Best Practice

At all times, your original data should remain untouched in a secure location; only work on a copy of your data. Start out by making some notes about which images are the best to use and make copies of those files in a local directory to work on. Using an image database, such as OMERO, can help because you can add ratings and tags to images to help you decide which images to use. Moreover, tools such as OMERO .figure allow the generation of figure panels in which the images are linked back to the database.

If possible, generate each figure panel using a script so that all panels have a uniform appearance. A scripting approach also means that a figure panel can be quickly regenerated if images need to be replaced. Assembling a huge multipanel figure using a script is not recommended, as it is simply more trouble than it is worth. Generating each panel automatically and then assembling the figure manually is the most efficient way to work. Use only with lossless image formats (i.e., TIFF) and only use compression—if at all—at the very final step.

Making Figures That Look Great

Tidiness and consistency are the key to making figures that look great. The ideal figure should have four clear edges (i.e., a rectangular layout) and any "white space" should be minimized. L-shaped layouts or figures with "ragged" edges look untidy, although sometimes it is avoidable (Fig. 7.2). For labeling, pick a font and use that one font for all panels in your figure and indeed for all figures in the document (paper, proposal, thesis, or report). Consistent sizing of fonts is also important. For labeling each panel (A, B, etc.), use, for example, 12 point and bold. Then only use nonbold and smaller point sizes for everything else. Journals set a lower limit for font point size in figures. Follow these and any other guidelines to save time later; they do not vary much from journal to journal.

Assemble your figure at the final size using a vector-based editing program. Vector format files are resolution-independent because lettering, lines, and other features are drawn automatically by the computer. Typical formats are PDF, SVG, and EPS. Image files are referred to as bitmap format. TIFF files are bitmap; other formats are PNG and compressed image formats such as JPEG. Adobe Illustrator is an ideal vector-based editing program for making figures, but it is expensive. Inkscape is a free cross-platform alternative that also has a command-line interface. Do not use Microsoft PowerPoint; it cannot make figures for scientific papers. Set the dots per inch (dpi) to 300 and set up an artboard as an A4- or U.S.-letter-sized page, whatever is appropriate. You can now add the components of the figure and scale them so that they look right and are not too big for a printed page. Although few journals still make a printed version, the PDF version of your paper is still based on hard-copy paper sizes.

The most difficult part of assembling a figure is correct sizing of each component. A typical resolution for display on a screen is 72 dpi. For printing, resolutions of 300 dpi or even 600–1200 dpi are typical of those required by publishers. This can cause confusion because the physical size of an image displayed on a monitor is quite removed from how it will appear on the page. Working with an artboard set at the final resolution and importing items at the correct resolution solves this problem. As you build your figure and add panels, make sure that everything looks OK. A typical mistake is that labels on axes of plots are too small (or rather that the plot has been

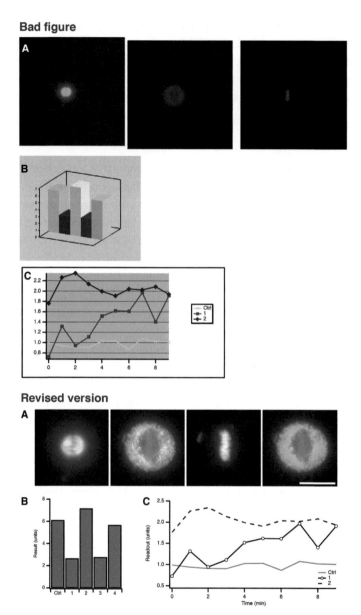

FIGURE 7.2. Bad figures and good figures.
Two figures put together from the same data. In the first attempt, there are many problems: (1) The layout is not rectangular, leaving lots of white space; (2) the cell images are not aligned, the cell is too small, and there is no scale bar; (3) lots of chart junk and unnecessary effects on the plots; (4) poor labeling of plots; and (5) inconsistent lettering. If these problems are addressed, the figure can be reorganized to look a lot better (*revised version*). Note that further improvements are still possible (e.g., better labeling of the cell images).

made at the wrong size). It is a good idea to take a break and return with fresh eyes to check that the relative sizes of all components are sensible. Another tip is to show the figure to someone else and get some feedback. This can reveal things that you no longer see because you have worked on it too closely. The other person should be able to understand what is going on in the figure without reading the legend.

The publisher may require the figures in a certain format and most programs can generate several file types. Be careful that any compression only happens right at the end of the process. Compression can be at the level of the file or in the image itself (pixels get resampled to reduce file size). This has the potential to make your carefully prepared figure look bad. If you are asked to supply a compressed version of your figures, test out a few different settings and always export a compressed copy rather than overwriting the original file.

Color Blindness

The inability to distinguish certain colors affects ~8% of men and ~0.5% of women. This causes a problem for the display of microscopy images, as we often rely on the overlay of red and green images to show coincidence of two signals, in yellow. You need to be aware that these images are meaningless for color-blind people. The overlay of two channels can be achieved using three pure RGB channels by duplicating the red channel in the blue channel to make a purple–green overlay. Now, coincidence of the two signals appears white, rather than yellow. The overlay and the distinction between channels can be seen by color-blind people. Ask yourself if a merge is really necessary if the results are already clear from the display of individual channels. Best practice is to show individual channels as well as a merge and to do so in gray scale. Gray scale is preferred to pseudocoloring, because the eye detects black-through-red differently to black-through-green or black-through-blue.

Contrast Adjustments

"Stretching" the contrast of an image is when displayed pixels have their value altered so that the full range of the LUT is used to display the image. This is necessary when there are only sparse very dark or very bright signals in the image. It is an accepted manipulation, but care is needed to make sure that clipping (a.k.a. saturation) does not occur. This is when bright features reach the maximum value (and dark features reach the minimum value), and detail becomes lost. If the image has dim features of interest, visualization can be improved by simply inverting the LUT.

The number of possible LUTs that can be applied to images is almost limitless. However, pseudocoloring is only useful for the display of data in a few scenarios. In the case where the distributions of two or more fluorophores need to be compared, an overlay can be made. In a simple case, one signal is placed into one channel (red) of an RGB image and the other signal is placed in another (green). Another scenario is where temporal or spatial information needs to be coded into the 2D image. For example, a microtubule plus-end moving across the cell can be presented in a single 2D image by using an alternating sequence of colors corresponding to each time frame. A 3D image might be represented in 2D by using a LUT to encode spatial information in the z-axis. Finally, proximity or density information can be represented using heat map techniques. Perceptually uniform LUTs designed for heat maps are now preferred rather than the older standard "rainbow" LUT. Any other uses of pseudocoloring are probably best restricted to making scientific eye candy for journal covers, rather than figures in a paper.

Transformation that follows a power-law (γ) correction means that both dim and bright features of interest in the same image can be visualized. As this is a nonlinear transformation, it has the potential to misrepresent the data.

Crops and Expansion

In most cases the original image will need to be cropped before it can be used in a figure. In the simple case, most images are initially captured as a rectangular file. Square images look better and are easier to work with, so you need to crop the image. The size of the ROI used to crop the images needs to be consistent and it might take a few tries to make sure that the ROI is sized correctly at the magnification used for all the cells you want to use for your figure.

Displaying a small ROI as an expanded view is a great way to show some detail while retaining an overview of the ROI in the context of the cell. This is best done using a macro in ImageJ before importing into the figure-assembly program. Attempting to do this manually in the figure-assembly software is a surefire way to make a mistake. A macro will ensure that the expansion is done isotropically and that the location of the ROI in the original image is indicated exactly. The location of the expanded view works well in an inset to the main image, but make sure it does not obscure any features in the original image.

Scale Bars

It is good practice to add a scale bar to micrographs, rather than stating magnification. This allows the reader to quickly assess the size of features in the image. For a picture of a single mammalian cell, a scale bar of 10 µm is useful because it approximately corresponds to the size of the nucleus. If a zoomed-in part of the cell is presented, a scale bar of 2 µm or 5 µm is more appropriate. For kymographs, indications of the scales in space and time are required. For a live-cell movie, or when presenting stills from such a movie, adding a time indicator for each frame is important. As with the crops and expansions and generation of the panel itself, do this programmatically before assembling the figure.

Movie Files

Journals tend to prefer movie files to be small. The best way to achieve this is to make the file physically smaller and also to use compression. Movie file formats are a minefield, but here is a simple guide to making a single channel movie file for publication:

1. Import your data into Fiji.
2. Set the contrast and crop to a square. Use *Selection > Specify* to minimize error.
3. Crop time appropriately (e.g., frames 1–108 of a 1–151 movie). Use *Duplicate* to do this.
4. Save a TIFF copy at this point so you can remake the movie if required.
5. Add annotation: scale bar (*Analyze > Tools > Scale Bar…*) and time stamps (*Image > Stacks > Time Stamper*).
6. If the cropped version is large (800 × 800 pixels), use *Image > Adjust > Size* to reduce the file twofold (400 × 400). Check *constrain aspect ratio* and *average when downsizing (bilinear)*.
7. Convert to 8-bit, not 8-bit color or RGB color.
8. Save as avi. I suggest 10 frames per second (fps) with no compression.

The avi file will be quite large, so you need to compress it. Use H.264 codec and a mov or other container, as specified by the journal. Using QuickTime on a Mac, this is as simple as opening the avi file and saving it again. The resulting file will be much smaller and ready to upload. Free tools such as HandBrake or FFmpeg will perform the same compression task. After generating the movie file, test whether it plays as expected. If possible, test on a Mac and a PC running the latest operating systems.

UNACCEPTABLE MANIPULATION IN FIGURES

You need to be aware of what is unacceptable manipulation of images. There is a long list of inappropriate things you can do that will misrepresent imaging data. They include erasing features, making collages of different cells into one image, or otherwise "beautifying" images.[37] This extends beyond microscopy. Notoriously, images of gels and blots that have had lanes spliced or the bands enhanced are a clear indicator of research misconduct and are the basis for retraction of scientific papers and the end of researchers' careers.

Use the minimum amount of processing of the data from the microscope to the image that you show in a figure. Simply crop your images and ensure contrast adjustments are equivalent between images. It is best to avoid γ correction or other enhancements unless you really know what you are doing. Always disclose any and all processing steps that have been made, and include these in figure legends.

There is no shame in showing data warts and all. Do not feel under any pressure to enhance your experimental data: It is what it is. Simply do the best experiments you can and obtain the highest-quality data that is possible.

The direction of travel in science is toward open data. At some journals, uncropped and unannotated blots must be shown in the Supplementary Information of the paper. At others, deposition of the entire data set is required. It is clear that in the future, sharing raw data and not just the "publication-ready" versions will become the norm. As a digital cell biologist, you will be ahead of the game.

◆ *Golden Rules*

- Aim to make figures that can be understood without reading the legend.
- Consistency and tidiness are key to great figures.
- Check the relative sizing and spacing of panels.
- Feedback is important. If something is not clear, you may need to add labels or make other changes. Ask for opinions on your work.

List of Software Featured in This Book

Task	Software	Link
Image analysis	ImageJ[6] Fiji[5]	https://imagej.nih.gov https://fiji.sc
Plug-ins and libraries	Bio-Formats[38] OMERO[39] TrackMate[18]	https://www.openmicroscopy.org/bio-formats/ https://www.openmicroscopy.org/omero/ https://imagej.net/TrackMate
Statistical computing	R[40] RStudio[7]	https://www.r-project.org https://www.rstudio.com
R packages	ggplot2[41] EBImage[42]	https://ggplot2.tidyverse.org https://doi.org/10.18129/B9.bioc.EBImage
Version control	git	https://git-scm.com
Figures	Inkscape Adobe Illustrator	https://inkscape.org https://adobe.com
Video files	FFmpeg HandBrake	https://www.ffmpeg.org https://handbrake.fr

Bibliography

1. Broman KW, Woo KH. 2018. Data organization in spreadsheets. *Am Statistician* **72:** 2–10. doi:10.1080/00031305.2017.1375989

2. Royle SJ. 2019. quantixed/TheDigitalCell: first complete code set, Apr. 2019. https://zenodo.org/record/2643411#.XLc3QC-ZMUE

3. Murre JM, Dros J. 2015. Replication and analysis of Ebbinghaus' forgetting curve. *PLoS One* **10:** e0120644. doi:10.1371/journal.pone.0120644

4. Williams E, Moore J, Li SW, Rustici G, Tarkowska A, Chessel A, Leo S, Antal B, Ferguson RK, Sarkans U, et al. 2017. The Image Data Resource: a bioimage data integration and publication platform. *Nat Methods* **14:** 775–781. doi:10.1038/nmeth.4326

5. Schindelin J, Arganda-Carreras I, Frise E, Kaynig V, Longair M, Pietzsch T, Preibisch S, Rueden C, Saalfeld S, Schmid B, et al. 2012. Fiji: an open-source platform for biological-image analysis. *Nat Methods* **9:** 676–682. doi:10.1038/nmeth.2019

6. Schneider CA, Rasband WS, Eliceiri KW. 2012. NIH Image to ImageJ: 25 years of image analysis. *Nat Methods* **9:** 671–675.

7. Team R. 2016. *RStudio: integrated development for R*. Boston, MA. http://www.rstudio.com/

8. Rueden CT, Schindelin J, Hiner MC, DeZonia BE, Walter AE, Arena ET, Eliceiri KW. 2017. ImageJ2: ImageJ for the next generation of scientific image data. *BMC Bioinformatics* **18:** 529. doi:10.1186/s12859-017-1934-z

9. Grolemund G, Wickham H. 2019. *R for data science.* https://r4ds.had.co.nz/

10. Chenouard N, Smal I, de Chaumont F, Maška M, Sbalzarini IF, Gong Y, Cardinale J, Carthel C, Coraluppi S, Winter M, et al. 2014. Objective comparison of particle tracking methods. *Nat Methods* **11:** 281–289. doi:10.1038/nmeth.2808

11. van Riel WE, Rai A, Bianchi S, Katrukha EA, Liu Q, Heck AJ, Hoogenraad CC, Steinmetz MO, Kapitein LC, Akhmanova A. 2017. Kinesin-4 KIF21b is a potent microtubule pausing factor. *Elife* **6:** 24746. doi:10.7554/eLife.24746

12. Applegate KT, Besson S, Matov A, Bagonis MH, Jaqaman K, Danuser G. 2011. plusTipTracker: quantitative image analysis software for the measurement of microtubule dynamics. *J Struct Biol* **176:** 168–184. doi:10.1016/j.jsb.2011.07.009

13. Aguet F, Antonescu CN, Mettlen M, Schmid SL, Danuser G. 2013. Advances in analysis of low signal-to-noise images link dynamin and AP2 to the functions of an endocytic checkpoint. *Dev Cell* **26:** 279–291. doi:10.1016/j.devcel.2013.06.019

14. Olziersky A-M, Smith CA, Burroughs N, McAinsh AD, Meraldi P. 2018. Mitotic

live-cell imaging at different timescales. *Methods Cell Biol* **145**: 1–27. doi:10.1016/bs.mcb.2018.03.009

15. Shen H, Nelson G, Kennedy S, Nelson D, Johnson J, Spiller D, White MR, Kell DB. 2006. Automatic tracking of biological cells and compartments using particle filters and active contours. *Chemometrics Intelligent Lab Syst* **82**: 276–282. doi:10.1016/j.chemolab.2005.07.007

16. Carpenter AE, Jones TR, Lamprecht MR, Clarke C, Kang IH, Friman O, Guertin DA, Chang JH, Lindquist RA, Moffat J, et al. 2006. CellProfiler: image analysis software for identifying and quantifying cell phenotypes. *Genome Biol* **7**: R100. doi:10.1186/gb-2006-7-10-r100

17. Chaumont FD, Dallongeville S, Olivo-Marin J. 2011. Icy: a new open-source community image processing software. In *2011 IEEE International Symposium on Biomedical Imaging: from nano to macro*, pp. 234–237. IEEE, Piscataway, NJ. doi:10.1109/ISBI.2011.5872395

18. Tinevez J-Y, Perry N, Schindelin J, Hoopes GM, Reynolds GD, Laplantine E, Bednarek SY, Shorte SL, Eliceiri KW. 2017. TrackMate: an open and extensible platform for single-particle tracking. *Methods* **115**: 80–90. doi:10.1016/j.ymeth.2016.09.016

19. Manders EMM, Verbeek FJ, Aten JA. 1993. Measurement of co-localization of objects in dual-colour confocal images. *J Microsc* **169**: 375–382. doi:10.1111/j.1365-2818.1993.tb03313.x

20. Zaritsky A, Obolski U, Gan Z, Reis CR, Kadlecova Z, Du Y, Schmid SL, Danuser G. 2017. Decoupling global biases and local interactions between cell biological variables. *Elife* **6**: 22323. doi:10.7554/eLife.22323

21. Lazic SE. 2016. *Experimental design for laboratory biologists: maximising information and improving reproducibility*. Cambridge University Press, London.

22. Lazic SE, Clarke-Williams CJ, Munafò MR. 2018. What exactly is 'N' in cell culture and animal experiments? *PLoS Biol* **16**: e2005282. doi:10.1371/journal.pbio.2005282

23. Colquhoun D. 2017. The reproducibility of research and the misinterpretation of p-values. *R Soc Open Sci* **4**: 171085. doi:10.1098/rsos.171085

24. Holmes S, Huber W. 2018. *Modern statistics for modern biology*. Cambridge University Press, London.

25. Motulsky H. 1995. *Intuitive biostatistics*. Oxford University Press, Oxford.

26. Tukey JW. 1977. *Exploratory data analysis*, Vol. 2. Addison-Wesley, Boston.

27. Goodman S. 2008. A dirty dozen: twelve p-value misconceptions. *Semin Hematol* **45**: 135–140. doi:10.1053/j.seminhematol.2008.04.003

28. Colquhoun D. 2014. An investigation of the false discovery rate and the misinterpretation of p-values. *R Soc Open Sci* **1**: 140216. doi:10.1098/rsos.140216

29. Sellke T, Bayarri MJ, Berger JO. 2001. Calibration of P values for testing precise null hypotheses. *Am Statistician* **55**: 62–71. doi:10.1198/000313001300339950

30. Benjamin DJ, Berger JO, Johannesson M, Nosek BA, Wagenmakers EJ, Berk R, Bollen KA, Brembs B, Brown L, Camerer C, et al. 2018. Redefine statistical significance. *Nat Human Behav* **2**: 6–10. doi:10.1038/s41562-017-0189-z

31. Sawilowsky SS. 2009. New effect size rules of thumb. *J Mod Appl Statistical Meth* **8:** 597–599. doi:10.22237/jmasm/1257035100

32. Ho J, Tumkaya T, Aryal S, Choi H, Claridge-Chang A. 2019. Moving beyond values: everyday data analysis with estimation graphics. *Nat Methods* **16:** 565–566. doi:10.1038/s41592-019-0470-3

33. Cumming G. 2014. The new statistics: why and how. *Psychol Sci* **25:** 7–29. doi:10.1177/0956797613504966

34. Weissgerber TL, Milic NM, Winham SJ, Garovic VD. 2015. Beyond bar and line graphs: time for a new data presentation paradigm. *PLoS Biol* **13:** e1002128. doi:10.1371/journal.pbio.1002128

35. Tufte ER. 1983. *The visual display of quantitative information*. Graphics Press, Cheshire, CT.

36. Brinton WC. 1915. Joint committee on standards for graphic presentation. *Publ Am Statistical Assoc* **14:** 790–797. doi:10.2307/2965153

37. Rossner M, Yamada KM. 2004. What's in a picture? The temptation of image manipulation. *J Cell Biol* **166:** 11–15. doi:10.1083/jcb.200406019

38. Linkert M, Rueden CT, Allan C, Burel J-M, Moore W, Patterson A, Loranger B, Moore J, Neves C, Macdonald D, et al. 2010. Metadata matters: access to image data in the real world. *J Cell Biol* **189:** 777–782. doi:10.1083/jcb.201004104

39. Allan C, Burel J-M, Moore J, Blackburn C, Linkert M, Loynton S, Macdonald D, Moore WJ, Neves C, Patterson A, et al. 2012. OMERO: flexible, model-driven data management for experimental biology. *Nat Methods* **9:** 245–253. doi:10.1038/nmeth.1896

40. R Core Team. 2018 *R: a language and environment for statistical computing*. R Foundation for Statistical Computing, Vienna, Austria. https://www.R-project.org/

41. Wickham H. 2016. *ggplot2: elegant graphics for data analysis*. Springer-Verlag, New York. http://ggplot2.org

42. Pau G, Fuchs F, Sklyar O, Boutros M, Huber W. 2010. EBImage—an R package for image processing with applications to cellular phenotypes. *Bioinformatics* **26:** 979–981. doi:10.1093/bioinformatics/btq046

Index